Epic Failures in DevSecOps

Volume 1

ISBN: 9781728806990
Imprint: DevSecOps Days Press

Publisher:
DevSecOps Days Press
48 Wall Street, 5th Floor
New York, NY 10005

www.devsecopsdays.com

Epic Failures in DevSecOps

Volume 1

Aubrey Stearn
Caroline Wong
Chetan Conikee
Chris Roberts
DJ Schleen
Edwin Kwan
Fabian Lim
Stefan Streichsbier

Mark Miller, Editor

"The stories presented here are not a roadmap. What they do is acknowledge failure as a part of the knowledge base of the DevSecOps Community." — Mark Miller, October 2018

Table of Content

Introduction

October 2018

We learn more from failures than we do from successes. When something goes as expected, we use that process as a mental template for future projects. Success actually stunts the learning process because we think we have established a successful pattern, even after just one instance of success. It is a flawed confirmation that "This is the correct way to do it", which has a tendency to morph into "This is the only way to do it."

Real learning comes through crisis.

If something goes wrong, horribly wrong, we have to scramble, experiment, hack, scream and taze our way through the process. Our minds flail for new ideas, are more willing to experiment, are more open to external input when we're in crisis mode.

The Genesis of an Idea

That's where the idea for this book came from. When I was in Singapore for DevSecOps Days 2018, Edwin Kwan, Stefan Streichsbier and DJ Schleen were swapping war stories over a couple of beers. The conclusion of their evening of telling tales was the desire to find a way to get those stories out to the community. They spoke with me about putting together a team of authors who would tell their own stories in the hope of helping the DevSecOps Community understand that failure is an option.

Yes. You read that right. Failure is an option.

Failure is part of the process of making the cultural and technological transformation that needs to happen in order to keep innovating. It is part of the journey to DevSecOps. The stories presented here aren't a roadmap. What they do is acknowledge failure as a part of the knowledge base of the DevSecOps Community.

What to Expect from this Book

This is the first in a series of books tracking changes and discoveries within the DevSecOps Community. The stories are by people who have been sloshing around in the swamps of software development for years, figuring out how things work, and most importantly, why things didn't work.

Chris Roberts starts us off with how the industry as a whole has failed us when it comes to software security. DJ Schleen, Edwin Kwan, Aubrey Stearn, Fabian Lim and Stefan Streichsbier provide a practitioner's view of being up to their waists in the muck of an epic failure. Caroline Wong and Chetan Conikee bring another view, peering into the murky waters of DevSecOps from a management perspective.

Each chapter follows a specific format:

- Overview, what were you trying to accomplish
- What went wrong, how bad was it
- How did the team try to resolve the issue
- What was the final outcome
- What were the lessons learned

Following this type of format, we should be able to create a series of stories, surfacing patterns we as a community can use to safely push the boundaries of software development.

Invitation to Tell Your Story

The days of stand-alone security teams isolated from the real process of development are coming to an end. Paraphrasing Caroline Wong, "Security needs to be invited to the party, not perceived as a goon standing at the front door denying admission". With DevSecOps, security is now part of the team.

After reading these stories, we hope you will realize you are not alone in your journey. Not only are you not alone, there are early adopters who have gone before you, not exactly "hacking a trail through the swamp", but at least marking the booby traps, putting flags next to the quick-sand pits and holding up a 'Dragons be here' sign at perilous cave openings.

On DevSecOpsDays.com, we'll be expanding the ideas and concepts talked about in this book. We look forward to your participation in the community, whether as organizers of regional DevSecOps Days events, as article contributors to DevSecOpsDays.com or as an author of your own Epic Failure on your journey through DevSecOps.

What would your warning sign say? We ask you to join our journey as we continue to learn from your Epic Failures.

Mark Miller
Founder and Editor in Chief, DevSecOpsDays.com
Co-Founder, All Day DevOps
Senior Storyteller, Sonatype

Chapter 1

We Are ALL Special Snowflakes

by Chris Roberts

Chapter 1
We Are ALL Special Snowflakes

Twenty-five years ago we, the network team and the various information security teams (in their infancy) walked into their CEOs' and CFOs' offices and proudly stated, "*We* need a firewall to protect us!" We started a chain of events that have led us to today's rather messy situation. For those 25 years or more we have continued to walk into the leadership's corner office and state that the next greatest thing will fix all the problems, secure all the things. We've done it by stating that the general reason we have to do this is because it's the users' fault, or the developers' fault, or the engineers' fault. Heck at one point I think I even blamed my grandmother for breaking security on the Internet.

For many years, we have continued to look at others as being culpable. We were special; we were the new warriors; the fighters of all things bad in the world; and, we were the only ones protecting the company against the perils of the modern era. Recently however, a growing number of folks have joined their voices together in earnest and started to rebel against the industry. Questions over tactics, over media portrayals and over the spread of fear, uncertainty and doubt throughout industry, in an effort to further the realm of security, are now being met with voices from within. Quite simply the tide is turning and our own industry is starting to introspectively look at what it's become and how it has to change to actually protect the very charges (companies and individuals) it has forgotten about, and to address all that is wrong within InfoSec/Cyber.

Fundamentally, we have realized how wrong we were and, unfortunately, how *very* wrong we still are. And, how much we have to learn and, more importantly, how quickly we have to learn it.

*Snowflake: an individual with an inflated sense of uniqueness, or an unwarranted sense of entitlement. (Wikipedia….)

A Closer Look At That History:

The primordial ooze of technology

Around 500 BCE, the abacus came into existence...and remained the definitive form of calculation until the middle of the 17th century. Think about that, over 2000 years with the same piece of technology in use, working effectively and not a single call center, vendor support network or venture capitalist in place to mess with it. The abacus saw us through some amazing changes in the world around us and, eventually, it was replaced by Blaise Pascal's mechanical calculator. The Frenchman's invention lasted 30 years until a German, Gottfried Wilhelm Leibniz, improved drastically upon the technology and gave us our first glimpse of what we know now as memory. Those of you who are paying attention during this quick history lesson will recognize our intrepid German mathematician/philosopher as the very same individual who presented the world with Binary.

One hundred years later (give or take a few) an Englishman, George Boole, took the whole concept of binary, threw it at a chalkboard and walked away a while later with an entire branch of mathematics that (thankfully) survives to this very day in the world we now have to untangle...Boolean algebra. The combination of binary and Boolean allows our modern systems to make sense, and simple decisions by comparing strings of ones and zeros.

So, at this point we have calculators, rather fantastic ones, yet still devices that require human input, and human hands at each stage. So, we've not (by definition) reached the age of computing which is where a machine is able to operate a program, or series of instructions, autonomously without the aid of a human. For this we have to look at the "father of computing" (no, not Al Gore) but Charles Babbage, and arguably the first "mother of computing" or computer programmer Englishwoman Augusta Ada King, the Countess of Lovelace (or Ada Lovelace). Between Babbage's innovative ways of looking at inputs, memory, processing, and outputs, and Lovelace's algorithms (Notes), as well as her forward thinking concepts about how these systems could evolve past simply number crunching, we see the start of the age of computers.

Now, we both fast forward to the 1880 and take a leaf out of the 1700s and combine the art of punched card with the technology of the era--the tabulator. Herman Hollerith, a statistician managed to take the census from a 7.5 year task to one of tallying it in six weeks, and full scale analysis in 2.5 years. The Tabulating Machine Corporation was set up (1896) which then changed names to the Computing Tabulating Recorder, and then to one that's familiar to us ALL in this industry, IBM.

The early days when we learned to talk *to* the technology

At this point, we had the machines, we'd worked out how to use them for some basic functionality, but we knew they could do more. Next, we had to work out *how*. For that we turn to another amazing figure in our history, Alan Turing, whom we can thank for the groundbreaking work on the theories of how computers process information. Turing is also known for several other key moments in our early history, that of the code-breaking machinery, the Enigma, and of the Turing test which is a method to see if a computer can be considered intelligent by measuring its ability to sustain a plausible conversation with a real human.

The time of Women (and why didn't we keep it that way for heaven's sakes!)

As we've already discussed, Ada Lovelace was the first programmer. Following her, we had the Pickering Harem, the rather ungracious name given to the female team that worked at the Harvard Observatory with Pickering processing astronomical data sets. The logic at the time was that work was considered clerical and the women could be hired at a fraction of the cost of a comparable male (incredibly, this battle is still being fought over 140 year later.)

The concept of the "human computer" harks back to these days and is often used as a reference as we move through history (NACA's computer pool being the 1930/40's version). Then we move to the 1940s and arguably the Grandmother of COBOL, Grace Hopper. She was the first person to develop and create a compiler. The simple logic being her belief that a programming language base on English was possible and that converting English to machine code, to be understood and executed by computers should be possible... Her achievements and her foundations led to the Mk1, the UNIVAC and host of other systems that have pioneered some of these countries greatest moments. While talking about the early days, it is well worth remembering the pioneering mathematicians, and their teachers and trainers who worked on the ENIAC computer: Adele Goldstine, Marlyn Meltzer, Betty Holberton, Kathleen Antonelli, Ruth Teitelbaum, Jean Bartik, and Frances Spence.

Fast-forward to the late 50s and early 60s and we run smack into the real-life history behind the movie, *Hidden Figures.* Dorothy Vaughan, Mary Jackson and Katherine Johnson among others who were literally the brains behind the NACA (later NASA) work at that time and went head-to-head with the likes of the IBM 7090s.

It's a man's world

Then we hit the 80s and computer science became "cool." (Let's face it, we were still nerds and geeks.) The computing focus started to shift towards it being a male profession. In part, we have to blame the advertisers, the game manufacturers, and the early PC developers--all of them targeted the male, the boys, and the companies that for the most part were run by "male geniuses". Within the college arena, computer science ended up in its own space, separated from the other sciences, humanities and other integrated areas, thereby reinforcing that separation from the rest of the baseline subjects that would typically attract a much more diverse crowd.

So, we have games for boys, computers being sold to the teenage boy, and advertising and college promotion aimed at the boys, and the whole field was having a massive influx of students, most of whom grew up with computers, who thought computing would make a good future. Not an ideal situation for fostering a diverse set of ideals, especially when there was a movement among some folks to treat knowledge as privilege or power, and should not be readily shared or used for the good of many.

The mainframe, we should never have let it out of the room

We had it all nicely under control, a room, a green screen and a person guarding the door--sometimes with a gun. You either got in, or got shot--simple binary response. The problem was the PC revolution was going nuts. Intel, Wozniak, Jobs, Kildall and Gates were all focused on bringing the computer to every home, every office and eventually moving the data from that monolith in the room to the desktop. For a while they co-existed, (remember the 3270 emulators and all the fun setting those up?) but eventually the PC market spawned the file server, the database server, and from that moment on the spread of data exploded.

We were doing so well and then--and this is the only time I'll say this--Apple ruined it all.

The distribution, time for tokens, and green screens

Token ring, MAU's CAU's (Mows and Cows) 4, 16, then this thing called Ethernet. In the middle was Banyan Vines and a host of other things. This was the time of "cable and re-cable"-- and do it all again with fiber-- oh, and now--Ethernet! The continual cracking open of the user's computer case for a new card, followed by a floppy drive install of drivers and you'd better hope you had the right settings otherwise the whole bloody thing fell around your ears.

Between changing out cards, computers, math coprocessors and installing WYSIWYG on the early spreadsheets, these were the days when we touted our knowledge, our experience and our absolute thirst to work out what was going on, how it all worked, *and* what we had to plan for next.

This was when we, the IT folks, should have been working much closer with businesses to understand *and* help them work out how they could and would use the technology. We didn't do a good enough job to see how the transition from the mainframe to the client server world would affect companies; we spent too long chasing the latest technology and not enough time listening to the very businesses we were beholden to.

The emergence of the giants, and the time we ALL wish we'd bought stock

This is the time we should have taken things seriously, taken a long look at the future and realized *this* was the time we had to make the necessary changes. We should have seen the shift in momentum and the emergence of the small companies with money behind them that took on and won against the giants.

We all kick ourselves for not buying Microsoft stock, or any of the other giants, that emerged in this era.

From BBS to Internet, the shift in momentum from hoarding data to seeing it *everywhere*

How many of us remember building, maintaining and using the banks of modems in various closets all across the planet? The boards, the early days of being able to share information that's turned into all the data everywhere. How many of us, back then, would have had the vision: that what we had, would eventually turn into what we now see all around us?

The proliferation of information is both fascinating and daunting. At least back in the early days you could, if you needed to, simply pull the plug on the board and a large chunk of things would go off line. These days that concept of being able to turn it all off has long since disappeared. We talk about being able to reset it should the worst happen, but at this point technology is *so* integrated into almost every part of our life that the negative consequences arguably outweigh any benefit. There's logic to that being part of the reason we simply accept the inevitable when it comes to the Internet and that identity theft, crime, and the complete lack of privacy is simply the price of pay to play.

I do not agree, I cannot agree and refuse to accept the current thinking. Quite simply, this indifference is something that has to change.

Apple's back and what the hell did they do with the "phone"

11 years ago, the iPhone was released and it quite simply changed the landscape, tore up the smartphone book and really kicked the mobile revolution into high gear. How revolutionary and how *much* did we mess up when it comes to being able to help this revolution be a safer and secure one? Let's explore:

First, the iPhone was an Internet device. It was a phone, yes, but when you look at the growth of data vs. voice traffic in the last 11 years, it quite simply moved the Internet from the desktop/laptop right into our hands *all* the time.

Second, we all became both consumers and proliferators of information. Back in 2011 it was estimated 400 billion digital photos were taken, fast forward to 2017 and the statistics are sitting at 1.2 trillion photos. We are simply moving everything we see, do and interact with into these devices that are sitting in our hands. *Unprotected.*

Third, we changed how we purchase software, applications and services. We use to spend time pondering the difference between all the separate software neatly stacked on the shelves. We pored through the PC magazine reviews, talked to the vendors and basically did enough due diligence to assume we'd made the best choice possible. Now we look at our $11 billion app-store shopping habits. Between the two main app stores out there, there are 6 million apps to choose from. Our diligence these days is limited to which one looks good, and answers these questions: Is it free? Does it offer in-app purchases? Can we get rid of the adverts? And will it integrate with whatever password manager we're using on the phone? We sometimes check reviews; however, we rarely care about whose hands are on the keyboard or what other data, access or "integration" they need to have with *our* device. Privacy and safety has taken a back seat to convenience and we, the IT/InfoSec/DevSecOps folks, have done little to help consumers understand the risks, nor helped mitigate those risks, until it's too late.

There are heaps more examples of how the iPhone has reshaped the world around us, and how we have adapted (not always in a positive manner) to its introduction into our lives. We have the computing power of a mainframe in our pockets, with the ability to change our lives in so many positive ways, yet we continue to fail to understand how best to use that. The introduction of the iPhone and its subsequent impact on the safety and security of how we interact with technology really drives home Lord Acton's advice in the mid-1800s that "absolute power corrupts absolutely".

All your technology, all the time, everywhere, with everyone

Somewhere in the middle of the mobile revolution, we arguably lost the battle. We were already fighting Bring Your Own Device (BYOD) and as IT and InfoSec, we threw up our hands, declared all mobile technology banished from our realms. And, we were summarily ignored by the users, businesses and the world in general. The mobile revolution moved the IT/InfoSec/DevSecOps teams from being the drivers into being the also-rans. Now, we had to adapt faster than we'd ever had to in prior years. We had to help a business understand how this technology would be used, and at the same time, deal with the implications of securing what was rapidly becoming a vanishing perimeter. There's an argument that when the laptop arrived we lost any vestiges of a perimeter, but for most of us who remember those early heavyweights they were about as "portable" as a desktop and as useful as a boat anchor. Because of those reasons, we still had some elements of perimeter because folks simply didn't want to have to deal with them. When the iPhone and subsequent smartphones arrived *any* perimeter quickly vanished.

Where Are We Today?

Let's take a quick, high-level look and break it down piece by piece

In 2017, across the whole information security industry we spent the best part of $90 billion; some of that was for the ongoing/running of existing systems; some of that was technical debt; and a chunk of it was for things that folks saw at conferences and were persuaded they needed to buy and integrate into their environments. At the same time we, as an industry, the protectors of our charges, managed to lose "somewhere" between 2 and 8 billion records--that's social security numbers, healthcare records, privacy information, banking/financial data and anything else that can be used against people to extort them.

So, how come we've managed to spend *so* much money and have *so* little to show for it? Why are we still looking around for the easy button and why the heck are we on track to spend even more in the next few years. All this as the criminal statistics are even more staggering. There's a consensus that our industry will provide continual fertile ground for criminal activities to the global tune of $6 trillion in anticipated damages in 2021, up from $3 trillion a few years ago.

Let's break it down into some quick manageable chunks and see what we can make of it:

Our Fragmentation

Our industry has fragmented, not just in the early days of IT when we split into networking, database, desktop, server and a small gathering of other areas (developers, etc.), but when information security overlaid itself onto each of the IT roles and exploded from there. We've been adding new and interesting titles each time a technology or buzzword is released. Today, we have hundreds of roles just within security.

Then, we overlaid the word "cyber" onto everything and that just confused everyone.

Then we formed chapters for ISSA, ISACA, ISC2, OWASP and host of other things.

And then we decided to have conferences, and those conferences spawned other conferences, which spawned "annoy the conference" conferences. Now we have a new one every week--which is good because it spreads the word—but bad because the word itself is too spread, out and diluted to the point of noise at times. And nobody *really* knows who to listen to, why to listen to them, *or* what logic to use to understand the value of what they are saying. So, we've taken a core group, fragmented it, expanded it, but have failed to retain any strong bonds between each of the fragments or any of the expansion kits.

Led by money not protection

"I've got an idea!" Both the greatest words to hear *and* the most frightening to those of us who have scars from being in the industry a while. Let me explain.

Your idea *might* be the next greatest, and safest mousetrap, but you have to develop it, market it, support it and critically tell everyone that it is the next best mousetrap. All this takes time--and critically money. So you borrow some money, friends, family and the kids down the street all chip in. You are beholden to them, so you don't sleep and you get the prototype out. Folks like it *but* you need to get to the market first, you need market share, you need to convince people that this *is* the mousetrap they need.

So, you borrow more money--this time from an institution and this time they want to make sure you are doing it right (their way, or with their help)--so they take some of your company and they help. Sometimes this is good, and sometimes this is a challenge, dependent upon who's doing the leading and who's doing the following. Meanwhile you need to still build the Mark 2 version and market it, *and* make it safe and secure, *and* you need to do it yesterday! And you still need to do the 101 other things necessary to run a business.

So, you go round in circles, possibly borrowing some more money from more people who want to help, and *now* you are beholden; you must make sure that those who have invested in you and your mousetrap get a good return. You put time and effort into making sure it's marketed, it's sold and it's "out there" and less time on the real reason for starting the whole process in the first place. The mousetrap has become simply a vehicle for making money, and *not* for protecting the very charges you set out to look after.

The illusion of red teams

"I want to be a penetration tester!" Congratulations! Join the queue and line up to break one of the 20-25 billion devices that will be in service by 2020/2021. How about we stop breaking things and spend more time fixing them? We're really good at coming in, breaking it and then wandering off all happy, full of ourselves that we've once again shown the developers, network types, systems folks or users that we can continue to break whatever's put in front of us. We'll even give you a nifty report (hopefully something more than a rebranded Nessus PDF.)

So, what's the solution? How about this approach: "I would like to work on defending and ensuring the integrity, safety and security of systems." This is far more collaborative with the entire organization, much more valuable--and given where technology is heading, and may result in much better long-term prospects.

Red is necessary. We need to be able to think as the attackers, to be able to maintain the security within the organization by continually testing the controls and technologies *and* the humans that protect it, but that team has to work in conjunction with the blue team, the internal defensive teams. Collaborative testing that engages on all levels has to be considered for the future.

Fool me once, shame on you, fool me twice shame on me: the plight of the auditor

I have empathy with auditors, quite a lot of it. I see how companies treat them, how they slap themselves on the back, congratulating each other that they fooled the auditors for yet another year. The auditor having once again failed to find all the skeletons in the closet, or simply didn't see the sleight of hand with documents, reports or whatever controls they asked for.

The marketing efforts, the million dollars spent on "look-at-me" booths

Walking around some of the more well-known conferences in the USA these past few years is depressing for more reasons than I care to note here, but for the sake of it, let's list a few:

- Look-at-me: the size and scale of some of the booths is obnoxious.
- Objectifying the women: we want more women *in* technology *not* as booth babes.
- The messaging: everyone seems to be able to fix everything, and their fix is the only one that'll do it.
- The pay-to-play keynotes: we want people to have earned that spot *not* bought it.

We have a LOT of growing up to do

It's been observed by folks far smarter than I am that this industry is unregulated. That should change. We hold life in our hands on a daily basis yet we have no formal training to do so. We hold the balance of the world's economies inside our systems, yet we have no formal background in how to do it best. We have access to intermodal, critical infrastructure and pretty much every facility we want to be able to get into, yet many of us have never stepped foot aboard a train, a cargo ship, a rail yard, coal fired plant or the innermost workings of a manufacturing plant. We have little to no direct experience or qualifications in the industries we are charged with maintaining, managing and ultimately ensuring the confidentiality, integrity and availability of.

We do this work, or have been doing this work, without any formal maturity within the organization, with minimal information flowing back to the business, with nary a glance in the direction of metrics, and with one hand on the wheel while juggling 101 other things (including the ever-increasing list of compliance questionnaires to fill out.)

We have to be *part* of a company, not special snowflakes

If we bask in our own unique talents, our own special gifts, we will be left behind. We can ill afford to continue down the path that we have been following. I do not want to be doing a follow up to this chapter in a few years time still pondering why we are blindly wandering around wondering why we've been left far behind by the very charges we should be protecting.

We know we have to come to the table, cap in hand. We have to come armed with humility and an understanding of the very organizations and entities we are protecting. We have to communicate in their language, and do so in a measured way where all parties understand risk, and how, as a single organization, to deal with it.

Those of us who consult with various companies also need to better understand our role from the beginning. Proffering advice and spewing statistics, basically blinding everyone with enough BS that we can grab the expenses check, and run for the hills will not work, should not work, and yet unfortunately, *has* worked in the past. Our role is to leave organizations in a better place than we found them. They put their trust and faith in us; the least we can do is honor that. We have failed in the past; we have to do better in the future.

Our own communities need to come together: DevSecOps

Everyone needs to stop blaming each other; everyone has to understand that we are all trying to do the right thing. The challenge is that we are not all pulling in the same direction. We have competing priorities; we have internal and external pressures, and we are not always in control of our own journey. If we can all pause for a moment, take stock of who we are as a community, realize that we function much better as a collaborative group. We can solve anything that's put in front of us and, if at the core of what we want is to simply make this a better place, then we should be able to find a common path, a common goal and start the "we" discussion and drop the "I" stuff.

I'll add in here that "we" means everyone of us, irrespective of race, color, creed, religion, sexual orientation, background, height, size, color or even if we wear kilts. The "we" has to be all of us, for a *lot* of reasons that go beyond the obvious ones of needing a diverse set of thoughts, considerations, approaches etc.

The momentum has to come from within; we have to fix ourselves

If we don't fix ourselves someone else will do it for us, and we probably won't like that. Let's not spend more time growling against whatever restraints have been put in place than actually accepting that we were the cause of the situation. The message here is clear: we're broken, and we know it. Let's fix ourselves rather than let some clown in the government try to do it for us.

What Do We Have To Learn?

We are still in our infancy, we are still being schooled by the very enterprises we're trying to protect, let alone connect. We should listen more and talk less. We have a lot to learn, but somehow we have managed to achieve what's never been done before in such a short timeframe. We have fundamentally changed HOW the entire planet works in a timeframe that spans one lifetime. The industrial revolution went from 1712 clear through to 1869 when the second revolution kicked off for an additional 44 years or so. During that time we went from steam to mass production of automotive transportation AND all things in-between. Conversely we've had computing power for about 80 years and have absolutely changed everything on the surface of this planet (almost without exception) our transportation, communication, food, health, shelter, etc.

So, in about one third to one half of the time, we've completely changed the surface of "us" but we've done so with some flaws in the whole scheme. We have taken on this task without a plan, we've been reactive and not proactive, fumbled a lot of what we could have done. In the last 30 years we have taken much of what was good and unfortunately left it behind in the pursuit of the almighty dollar (or whichever currency you are sitting in.)

So, we DO have a lot to learn, let's take a closer look at some of those things:

Comms

Communications--this is all encompassing, between the technical teams, between each other, to the users, managers, business, humans in general and especially between each of those bloody applications we keep pumping out.

Borrowing something from the healthcare field

A simple question to ponder on that could have some far reaching consequences. Would our industry learn from a simple statement of "First, do no harm".

Measuring Everything!

Metrics--we don't know how we're doing (apart from the fact we know we're doing poorly). We rarely are able to accurately tell people how things are going and our ability to accurately predict our progress is scary beyond belief. If we were a bank, we'd be rounding out our accounts and crossing our fingers, and we've rounded to the nearest "0"!

Stand together or fall alone

We all have to come together as a collective. Information security is a family, albeit a dysfunctional one at times, but still a family and we have to do a better job of acting like one. We that would be government, civilian and military *must* come together. I see too much wasted effort, duplicated effort and simply crossed paths that prevent us from being effective.

"I" will fail. "We" will succeed

This is simple, the message says it all. "I" can't do this alone, that's the "I" that looks back at you in the mirror in the morning, or the "I" that gets a cup of tea or coffee to start the day. It's the "I" that sits in meetings wondering *how* to fix things. *That* "I" is not going to be able to do it alone. "We" have to come together to do this in ways that are collaborative, effective, and essential to our future.

How Do We Do It?

Congratulations, you've made it this far. This is far from light reading, and as an introspective look at the very industry that I've been a part of for a long time, it's a rough read. However, all is not lost (famous last words). The following area breaks down some of the thoughts scattered in the earlier paragraphs.

Some basics that should help each one of us

- Security and safety are *not* afterthoughts; we should work out how to communicate these effectively across all areas, personal and professional.
- Safety will resonate much more effectively if you can cohesively use it in place of securing "everything" The concept of that very iPhone being a safety concern is likely to resonate more than simply waggling the finger under someone's nose because they still use 1234 to unlock it.
- Build safety and security in from the very start of a project!
  - Build it like your *mother* is going to have to use it
  - Built it as if attackers *are* going to come and *tear* it to shreds because they will.
  - Build it with insight and foresight: this is your baby, don't make it ugly
- Help everyone on the project, educate and advise them:
  - Show them pictures of your mother when it comes to user interfaces and more passwords
  - Show them pictures of "forensic files" when it comes to handing credentials etc.
- Use *all* the resources at your disposal to make something good.
- Make it adaptive and predictive. Make it preventative. Don't make it reactive; remember evolution is good, look at the future and build to that.
- Safety and security have to be a mindset.
- Safety and security have to be the differentiators.

    - Your organizations actually might thank you!
    - Your customers *will* thank you!
    - Use it to your advantage in marketing.

- Vendors need to be held responsible for delivering safe and secure products to *all* their clients *all* the time--not 3 years down the road *if* enough people scream.
- Integrators need to be held responsible for educating partners *and* vendors *and* choosing wisely.
- Feel like we are flogging a dead horse? But wouldn't it be nice for once to be unable to break into a company because defaults or outright dumb passwords had *not* been used or tolerated.

So, there's some baseline points to build from, something to consider next time a project kicks off or a vendor comes round or the leadership team asks for input. I hope this helps, I hope this starts the very REAL discussion that needs to happen because if not that tsunami of technology IS going to drown us all.

Why Us? With knowledge comes responsibility.

This is not something that we can leave to others. We created the mess, and we have to fix it with the help from the younger generations coming into this industry and the others in the general business population, and yes that means everyone! Blue-collar, white-collar, no-collar, Gen X, Y, Z, A, Millennial etc. You get the idea; we have to think outside of our comfort zone.

Some Final Thoughts:

Some final contemplation on what the future holds AND why change has to happen…

Technology and the edge of the cliff

Around 248 million years ago the first dinosaurs appeared, and for the next 183 million years Mother nature nurtured and grew an entire planet worth of stuff, up to and including shifting continents around to ensure that the right species got to the beach at the right time. However in all those years, never once did Mother Nature deem it necessary to give the Tyrannosaurus Rex thumbs, or any means by which to successfully use a knife and fork. Think about that for a moment, 183 million years and the best that could be done was cockroaches and crocodiles. Then the reset button was hit, it went quiet for a while and we came along. 200,000 years ago we really started to kick off (after coming out of the trees 6 million years earlier) and 12,000 years ago we stopped hitting each other with bones and started on our quest for knowledge.

Today we've not only got our opposable thumbs working overtime on a multitude of pocket devices, we're evolving our bodies and minds to a point where even Mother Nature's not gotten a map--and that's the problem. We've lost the plan. Our species evolved faster and with more flaws than Mother Nature's SDLC had planned. Now we've thrown away the designs, cast out the integration and testing, and are doing our very best to head over the cliff at full speed without a care in the world.

Arguably, our role is to change that, to take back some of the technical control, to reapply a lifecycle change management and to better understand the impacts of what we're doing, who we really are, where we're going *and* how we'll get there.

Artificial Intelligence wakes up…

In 1949, George Orwell introduced us to the dystopian future of 1984 in which independent thinking and individualism were ground out of our society. Ironically enough in 1984 we were introduced to the means by which such individualism would eventually be our undoing: the machines. In this instance, a 6'2" Schwarzenegger was sent back from 2029 by a machine that gained consciousness in 1997. If you are still with us insofar as timelines, (believe us this is just ONE plotline) we've apparently been persecuted by machine for about the last 30+ years and we are yet to realize it.

So, the questions are simple and we'll have to address them soon enough:

- Will the machines wake up?
- Will they resemble us, need coffee, be grumpy before 9am, demand breaks and sulk when told "No!"
- Will they take one look at humanity and wonder HOW the hell we've survived to this point?
- Will they take the steering wheel away from us, throw us in the back of the car and take over?
- Will they consider us nothing more than a pest and deal with us accordingly?
- Will they take one look, realize we're a lost cause and head for the stars?
- Will they work with us? Will we listen? Will we have a choice?

OR

- Are we barking up the wrong tree? Will we simply evolve beyond the separation of human/machine and integrate ourselves?
- Will we take a different path and revert to simply being signals and integrate at a conscious/electron level?
- Lets face it, this shell we occupy is fragile and temporary in nature. Can we simply leave it? What *is* human?
- To these points, we are going to have to seriously look at the following:
- Whose hands are on the keyboards, how influential is that in the overall design?

- Whose countries are at the forefront of design and what implication does that have?
- Who is paying for all this and what are those implications?
- What is privacy and do we need it? Can we have privacy and *actual* artificial intelligence?
- How do we account for all 7.4 billion of us on this planet when we are designing a system to think for us?
- What happens when the system decides to restore from a backup. Which one is the "true" system and which one is going to suffer from an identity crisis?

Biotechnology and Nanotechnology:

The barrier between humans and computers has been chipped away for many years; however, we've now crossed into territory that goes beyond embedded technology, chip placement or prosthetics. We are at a point in evolution where our living breathing bodies are directly interacting with the very systems we design through the continued evolution in biotech and nanotechnology development. The upsides of these breakthrough in Micro/Nanodesign are to be celebrated; however, with all good things comes the respect that needs to be shown to the invasive and communicative nature of the solutions. This is where we have taken a long hard look at the proposed architectures, and over the last few years demonstrate some of the challenges in the communications and security around letting computers loose in the bloodstream.

In reviewing the current security and communications of nanosensors, nanoantennas and other technology and the associated architectures we find that once again we're heading off the technology cliff at full speed with nary a glance behind at the safety and security implications. The fact that we can hack the human with nothing more than a modified BladeRF/HackRF setup should be pause for concern, yet the industry charges ahead oblivious to anything more than the advancement of human/technology integration.

Consciousness and the exploration into the simple fact we might be nothing more than a soggy walking bag of electrical sparks...

Taking security through cognitive analysis to the next level.

We are who we are; each of us *is* unique in the manner we have arrived at. That *is* something that can't be taken from us.

- Influencers--consider this the nurture side of things.
- Surroundings, what around me is helping to determine what/who I am and what I am doing?
- My life and I (Mother Nature started the process and we've been tuning ever since) The processes that have taken us from inception forwards, each of us has a unique "life" that is particular to us and can be recalled (depending upon what/when) at will and without any external influence.
- Given this logic and the work that's being undertaken in the lab to penetrate the brain through a neural engineered system that takes the neurochemical signals our brain produces, turns them into binary and then transmits them to and from a secondary device through NFC and some other tools.
- The logic here is that we are now at a point where we can both detect signals from the brain as well as implant/sense millions of signals coming to/from it through various means and methods (DARPA has several projects on neural interfaces etc.)
- The other option here is that we have the ability to detect weak electrical fields in the brain. We can detect and translate those waves in the field into bits/bytes; from there, we look to turn this into machine usable language.
- We would have a unique identifier that the computer can relate to. It would identify when we purchased it, how we configured it, when we used it and (if in a corporate environment) when it was assigned to us *and* what/who we are and how we should be using the system. There will be no need for us to have passcode, passwords or anything as archaic as actually writing down the access permissions that we need.
- From a validation and acceptance standpoint we would be able to provide a unique history of who we are, and what were our

interactions, influences and other deciding factors that make "us". Those criteria would provide the necessary collateral for the systems to communicate, realize access should be provided and then simply move on. The upside of this is we could provide an almost infinite number of criteria based on our experiences that would allow for a unique interaction/key exchange every time we needed to interact with a controlled system.

- The concept here is to develop the device, the interface and the architecture necessary to be able to support the unique identifiers that are "us" in such away that they don't need to be stored on any device that isn't "us". The computer, phone, IoT, car and other devices requiring validation (software, web, cloud and others) would be able to interface with "us" in a manner that is both reminiscent of a one-time-use pad (think of the unique combinations each of us has insofar as memories etc.) combined with the access controls unique to the neural network that we'd be monitoring.
- The ability to read the digital patterns is being developed both at an intrusive and non-intrusive level. There would be some logic flow on which is more relevant/opportunistic. Logic says non-intrusive *but* with chips being implanted, the ability to use micro-antenna for receipt/send capabilities is simple. The digital signal is read from the brain based on either current micro-electronic signal inputs or two other methods that are sitting on a whiteboard. From this point, it's a matter of identifying "us" and facilitating the necessary handshake with the endpoint. There's no digital signature, no digital passport, no use of DNA or anything that can be compromised, it's "us" nothing more, nothing less. The signal and the memory processes change on a constant basis BUT can be keyed into certain signals based on key events that would be synchronized between the "us" and the endpoints.
- The programming or imprinting of the endpoint devices would also be unique. The memory of them and of using them and obtaining/first use etc. would be encoded. At that point, it's simply a matter of human recall to ensure the correct handshake--nothing more.
- So, you get the idea, this is not only looking at the future, it's actually eating my own dog food. We (as an industry) have spent 25 years or more screaming at the top of our lungs about passwords, and this is one method to simply do away with them, no Band-Aid, no patching, no excuses or blinky lights, no bullshit, just a

way to fundamentally remove one of the worst barriers we have had to deal with.

- And, while I'm at it, I'm training a neural network on a separate machine to understand "how" I'm reasoning certain situations and letting it work on predicting outcomes. So far it's got a good set of baselines, understandings *and* situational awareness parameters among other things and is sitting at about 75% accuracy.

In Closing:

So, there you have it, information technology, security and all things cyber laid bare. The ugly truths exposed and in the middle of the book, we find something that looks like a rather large pile of poo that someone's got to clean up. It is a simple truth that we have failed the very charges we were meant to be looking after. It's excusable that it might have taken us a few years to realize what the criminals were up to. It might have taken us until we got past Y2K and heaved a sigh of relief that the following day actually happened. However, it is simply inexcusable that an industry and a field that has so many resources at its disposal continues to fail so spectacularly. You want an example of epic failure; take a look in the mirror. You want to fix the bloody mess, take another look in that same mirror, heave a heavy sigh, get your arse in gear, snowflake, and buck your bloody ideas up. ***We*** are all the solution; let ***us*** collaborate!

Footnotes (thanks to Mr. Pratchett for the inspiration!)

- Firstly, thank you to Mark Miller and the team behind this. There is NO way I would have undertaken anything like this on my own. All credit to him for having the faith that I'd actually be able to get things to him in time (almost-ish).
- Secondly, HUGE thanks to Johanna for the editing, suggestions, and overall crafting at the twelfth hour!
- I realized the second day I hit this that I was not able to type correctly, came to work out I can't type with acrylic fingernails… so off they came, and less mistakes, more productivity and better language directed AT the computer.
- Comparing the industrial revolution to our world brought flashbacks of having to sit in school and learn about trains and Isambard Kingdom Brunel…that's probably another deep-seated reason I hate trains and hack them whenever I can.
- Music listened to while writing this: Audiomachine, Thomas Bergersen, Led Zeppelin, Hans Zimmer, Epica, Two Steps From Hell, Queen, Brand X, and, Iron Maiden.
- Having to introspectively look at our industry through this lens hurt. I spent more time wondering "if" we can recover than I want to admit. It has made me more determined to fight the mess and walk shoulder to shoulder with anyone else who's going to be part of this movement.
- The fact I can use primordial ooze of technology makes me grin…
- I DO want to point out that it is officially 5 hours past the deadline, and apparently in 6 hours time the reviewers get access… and I'm still sitting here with a good single malt and munchies working.
- The reference to absolute power and its ability to corrupt is a personal frustration that I have with the whole use of technology. We have at our fingertips some of the most amazing tools that could do so much *good* in this world, could help to solve so many problems, yet we spend so much time wrapped up in them in *so* many meaningless ways. Instead of helping society, they have become the worst ever time sinks yet developed.

References:

http://www.softschools.com/timelines/computer_history_timeline/20/

https://www.ducksters.com/history/us_1800s/timeline_industrial_revolution.php

https://www.warren.senate.gov/imo/media/doc/2018.09.06%20GAO%20Equifax%20report.pdf

https://www.explainthatstuff.com/historyofcomputers.html

https://en.wikipedia.org/wiki/Women_in_computing

https://cybersecurityventures.com/hackerpocalypse-original-cybercrime-report-2016/

https://www.csoonline.com/article/3153707/security/top-5-cybersecurity-facts-figures-and-statistics.html

https://www.goodcall.com/news/women-in-computer-science-09821

About Chris Roberts

Chris currently works at Lares; he's the chap doing adversarial research and other things. Prior to that, he's founded or worked with a number of companies specializing in DarkNet research, intelligence gathering, cryptography, deception technologies, and providers of security services and threat intelligence.

Since the late 90s, Chris has been deeply involved with security R&D, consulting, and advisory services in his quest to protect and defend businesses and individuals against cyber attack.

As one of the well-known hackers and researchers, Chris is routinely invited to speak at industry conferences. CNN, The Washington Post, WIRED, Business Insider, USA Today, Forbes, Newsweek, BBC News, Wall Street Journal, and numerous others.

And worst case, to jog the memory, Chris was the researcher who gained global attention in 2015 for demonstrating the linkage between various aviation systems, both on the ground and while in the air that could have allowed the exploitation of attacks against flight control system.

Chapter 2

The Security Person Who Is Not Invited Into the Room

by Caroline Wong

Chapter 2
The Security Person Who Is Not Invited Into the Room

I am going to tell you about what it is like to be the security person who is not invited into the room. My name is Caroline Wong, and I am currently the Chief Security Strategist for a penetration testing as a service company called Cobalt.io, based in San Francisco.

I started my security career 13 years ago, leading security teams at eBay and Zynga. These were super cool places to begin working in cybersecurity. In both cases, we were running online operations 24x7 with millions of simultaneous users daily.

eBay had an uptime requirement of 99.94% and as one of the first major electronic commerce shops, enabled strangers to transact with each other over the internet.

Zynga was growing incredibly rapidly as an early adopter of Amazon AWS. In 2009, the Zynga game Farmville launched and in just a few weeks, the game went from zero to 10 million daily active users. A few months later, it rose to 80 million daily active users. We also had some incredible data stores. One game logged more than 30 billion transactions a day!

Why Does Security Matter for DevOps?

So in these type of environments, why does security matter?

At eBay, a public company handling payments between customers, PCI and SOX compliance were big initial drivers. In Zynga's case, we were also getting ready to go public. An IPO, of course, means that you have to be SOX compliant.

For many of today's DevOps companies, security is also critical to establishing trust between organizations in an "easy come, easy go" SaaS environment. Corporate and enterprise companies demand proof of security practices and technical assessment results (such as manual penetration test reports) to demonstrate that the software products and services they are using meet their standards and requirements. Vendor security assessments are becoming a regular part of the software procurement process and the due diligence that a company conducts when they are considering a merger or acquisition.

Of course, no one wants to see their company's name in the headlines due to a security breach. No one wants a security breach, period.

Pretty good reasons for DevOps to have security, *amirite*?

Head Banging Moment: Electronic Commerce

For application security at eBay, we were doing a ton of defect discovery. This involved lots of work finding bugs through penetration testing and vulnerability scanning, and getting information from external security researchers via responsible disclosure.

It seemed like every week we would go to the development teams and say, "Here is a pile of bugs. It's super important that you fix these. Right now."

They would basically close the door in our faces. Pretty soon they stopped showing up to our meetings. Which is fair, really - we were just giving them extra work to do.

For cyber security professional, this might sound very familiar. It's really a bummer, because it can feel like you're not really making progress at work. We were finding all sorts of security issues, but that's only *half* of the solution.

In order to actually improve the security posture of software, you've got to find security problems and risks, *and* you've got to fix or mitigate them.

Head Banging Moment: Online Gaming

At Zynga we thought, "Great, the company is getting ready to go public."

We assumed that this meant that we had a big stick that we could use to help get our security work done. So I took the NIST 800-53 security standard, which happens to be 387 pages long, and I customized it for Zynga. I condensed it down to just around 50 pages. And then I started trying to set up meetings with technology stakeholders to buy into our policy.

Guess what? Nobody showed up to our meetings...again.

So, we tried something different. The CISO of Zynga at the time said "Well, we need people to understand that information security is not only our job, it's their job too. Let's make a roles and responsibilities matrix." So we made this gorgeous RASCI matrix which stated what everyone is supposed to do.

I am pretty sure that no one read it. Once again, as a security professional, it was super frustrating.

I found myself constantly wondering, "Why doesn't anyone seem to *care* about security?"

Path To Epiphany

I care about security, and I felt like everybody else should too. At the time, I couldn't figure out why our interests were not aligned. My path to epiphany was a series of learning moments over time.

The first thing that we started to try and do was to ask some questions that we had not asked before. Specifically, we began to ask the developers and the technology teams, "What is important to you? And what are you trying to accomplish?"

It turns out that developers have important things on their minds, like quarterly goals and deadlines. They are trying to build new features so they can make money for the business and hit their personal and team targets.

When we approached them with piles of security work to do, it did not instill trust, and it risked putting them behind schedule.

What we needed to do was to put some context around the information we were sharing and the requests we were making. I actually applied many different principles from **Shannon Lietz's DevSecOps Manifesto**.

Data and Security Science over Fear, Uncertainty, and Doubt (FUD)

At eBay, we teamed up – the security team and the development teams – to define a measurable objective for our common goal.

Conveniently, the CTO had just approached our CISO and asked for a security score for each of the applications on the customer facing websites. This was extremely convenient, because his question put us in a position to ask ourselves, "Well, what should that score be?"

We decided that for every customer facing website, it was going to be a **defect density** score. We wrote down the total number of security bugs for each application and divided it by MLOC (million lines of code) to normalize the score across more than a dozen different applications.

Once we had buy-in from the decision-makers, our application security team approached the developers and said, "Historically, and realistically, what kind of bandwidth do you have to address and remediate security vulnerabilities?"

At the working level, we decided we were going to go for a 20% reduction in the number of vulnerabilities on the customer facing websites over a period of one calendar year.

We tracked and reported the numbers every month to the developer teams, the CTO, and the CISO. By the end of the calendar year, we had achieved our common goal.

Note: 20% is not a number that security people usually like. Security people like a number like 90%. Or 95%. The thing is, if we had gone and said "we're going to try and eliminate 90% of the bugs on the website," we probably would have gotten the same response that we got before -- development teams would stop inviting us to their meetings, they would stop coming to our meetings, and they might even stop reading our emails.

Business-Driven Security Scores over Rubber-Stamp Security

We took this idea a little further when I was on the Zynga security team, choosing to follow another one of the DevSecOps Manifesto principles.

Instead of going to the developer teams and saying to them, "Here is a big pile of bugs," we would say: "Based on conversations that we've had with you, about your business, the architecture of your application, and how it works, we have created for you a Studio Risk Profile."

We then presented a chart showing the bugs in a visual format. The Y-axis showed "bug severity" and the X-axis showed "value to attackers." In this way, we were able to leverage the information from our threat modeling exercises to prioritize bugs against each other.

In one example, a security vulnerability that allowed a player to cheat (perhaps by duplicating an in-game asset) would have been a relatively low priority. On the other hand, a security vulnerability that allowed one player to steal an asset from another player would have been a higher priority. A bug in Exampleville, the game created by the Shared Technology Group that was used to build all the rest of the games, was a very high priority due to the multiplying effect of duplicating that code in multiple places. In some cases, security vulnerabilities and how players used them to cheat and manipulate games were demonstrated in user group forums and on YouTube. If this type of information was publicly available, that was another good reason for the security team to increase the priority of the bug fixes. Finally, certain games had active secondary markets that were associated with the exchange of in-game goods. We focused more attention and more resources on fixing the security vulnerabilities that could lead to larger impacts on secondary markets.

We used threat modeling to help us expand on the idea that not *all applications are created equal, not all games are created equal, and not all bugs are created equal.*

There was a sense of risk ranking and associated risk-based security controls and priorities that were bestowed on each found security vulnerability, depending on a number of different business context-driven criteria.

This enabled the security team to prioritize bug fixes for developers, and increased trust between the teams.

Open Contribution and Collaboration over Security Only Requirements

It's a really different approach to go into a room and say, "Listen to what I have to say. This is what I need you to do." versus going into a room and saying, "Here's a problem that we need to work together to solve. Here's how I think we should approach it as a group."

In the past, I had taken a 387 page policy document and modified it down to a 50 page policy document. I then tried to shove it down people's throats.

My new approach involved hosting a policy building workshop, inviting stakeholder teams - Legal, HR, IT, Operations, Networking - and saying to them, "We're going to go public, and that means we need to be SOX compliant. Here's what we need to do. Let's talk about it and prioritize together. From a security perspective, here's the bar that we need to meet. *You* tell me how we're going to go about it."

We actually ended up with security policies that were signed and dated by the teams who had written them and who were expected to adhere to them. This made it such that they were going to follow the policy. #TeamworkFTW

Sharing Threat Intelligence versus Keeping Information to Ourselves

Prior to my current role at Cobalt.io, I was a management consultant with a company called Cigital (now Synopsys). I delivered more than three dozen BSIMM assessments. BSIMM is a software security framework. If you're not familiar with it, I highly recommend checking it out at www.bsimm.com.

There's an activity in BSIMM with regards to sharing threat intelligence. During my team conducting BSIMM assessments, I talked to more than 36 organizations doing software security, and I asked if they shared threat intelligence throughout their organizations. The vast majority of teams do not.

At Zynga, we decided that in order to build trust with the other teams at the company, we were going to tell people when something was up.

For example, if our executive staff began to receive malicious email attachments, not only did we have the technology in place (Ironport) to strip out those malicious attachments, we also sent messages to them letting them know that they were being targeted. We wanted to make sure they knew what was happening.

Leading up to the IPO, there were also many recon attempts on Zynga employees who were learning about the IPO and what it meant for their stock ownership. We emailed everyone to make them aware of potential social engineering techniques to help them learn about which messages were trustworthy and legitimate, and which were not.

The primary platform where Zynga games could be played at the time was on Facebook. So, all Zynga employees were fairly active on the social media network. Every once in awhile, Zynga employees and players would receive friend requests from an account pretending to be Zynga Security. These were fake, and we alerted the employees and customers to the threat so that they would not fall for it and mistakenly connect with and share any information with the malicious account creator(s).

I believe that by being transparent about what was going on and not holding that information so close, we were actually able to have better relationships with the other folks at the company.

In Conclusion

Security folks sometimes get a bad rep for being the team that always says no, or coming to the table with work for others to do without business prioritization or justification. That doesn't always have to be the case.

In a DevOps world where software drives revenue, secure software protects revenue. Security goes from being a cost center to a business driver.

It's critical that security professionals embrace an approach that is curious about other teams and the business. It's only by partnering with others that we can secure the technology that we build, buy, sell, and operate.

About Caroline Wong

Caroline is a dynamic cybersecurity expert with more than a decade of industry experience as a day-to-day manager at eBay and Zynga, product manager at Symantec, and managing consultant at Cigital. These days she helps connect DevOps companies who want to improve their cybersecurity posture with hackers who can help find their problems before the bad guys do.

Caroline received a 2010 Women of Influence Award in the One to Watch category and authored the popular textbook Security Metrics: A Beginner's Guide, published by McGraw-Hill in 2011. She graduated from U.C. Berkeley with a B.S. in Electrical Engineering and Computer Sciences.

Chapter 3

The Problem with Success

by DJ Schleen

Chapter 3
The Problem with Success

Introductory Overview

After months of planning, we finally added a security control into our automated build pipeline. This particular integration was the first of many we had planned to roll out over the next few years. Static Analysis and Security Testing (SAST) was ready to check our source code for security vulnerabilities. Little did we know that we just set up the security team to experience years of frustration.

First Baby Steps

Back in the day when web development was the new thing, I remember sitting beside a product owner and pushing to production as soon as I made a code change. Sometimes the product owner would turn and say to me "I think that's not working - we should probably change it". I'd make the revision and push it to production right away. Now that wasn't DevSecOps or even DevOps for that matter; it was the 1990s and was what we now call *Cowboy Coding.*

I had long forgotten about this memory when I walked into a small Denver startup six years ago for my first day of work. I was told that before the end of the day I would be pushing code to production through an automated process. I had never heard of this kind of thing before and had never seen any kind of leadership support this idea in the past. It seemed like Cowboy Coding but was the furthest thing from it. It was the early days of DevOps, and my first look into a new way of doing things.

About a year later, I approached the CTO and asked why there weren't any security controls in our automated deployment process. We were managing Personally Identifiable Information (PII) for millions of people as well as their relationships with each other. I was told that "*security wasn't a priority at the organization at this time*". I was floored. I should have anticipated this response at a twenty-five person startup, but the answer was completely unacceptable.

Frustrated by the lack of responsibility the startup had for managing their customer's information, I decided to leave. I had learned so much about automation and rapid deployment of software, but the discomfort I felt about the lack of security wouldn't go away.

For the next six months I worked as an independent consultant. I focused on Penetration Testing and Ethical Hacking, but it was Social Engineering and Red Teaming that I enjoyed the most during that time. When I was approached by a prior coworker about a job opportunity at a small company recently acquired by a Fortune 50 organization, I decided to see if the position was a fit for me. Don't get me wrong, Red Teaming was a great experience and a ton of fun, but breaking into buildings, dumpster diving, tailgating, and deception started to wear on me after a while. I was also away from my family way more than I wanted to be.

I submitted my resume and was contacted to begin the interview process. At the suggestion of the IT leadership, I was invited to attend a large release planning session in their developer area. To my surprise, the room was filled with every employee. It didn't matter what their role was; there were people from Human Resources, the Design Department, the Executives, and every Developer, Tester or Operations employee at the company. This *Release Fair*, as they would call it, was when all of the feature teams would create a large planning board and share it with each other. The board described what features the team would be developing in the next twelve weeks.

I didn't pick up on it until many years later, but the Release Fair I had just witnessed was a grand display of *Culture* - the most important Principle of DevSecOps.

I took the job.

A New Way of Working

After coming on board I learned that releases were organized into six iterations - the first five were development iterations, and the last was a Innovation, and Planning Sprint. Everyone looked forward to this last iteration as it was a time where we could all take a deep breath and discuss the previous release.

There was a bigger *purpose* to this sprint for us; it was a two-week allocated time span where developers could harden the code they delivered over the course of the Sprint. Of more importance was that this iteration provided a time where we could use our imagination and develop whatever product or proof of concept *we* wanted to. Often these projects would contain functionality that impressed the Product Managers so much that they would add them into their product roadmap. It was a pretty impressive way to build software products.

My journey at the company began on the iOS team where I had a number of responsibilities. Not only was I a mobile developer working on the company's flagship product, but one of my more challenging assignments was to figure out a mobile application delivery pipeline. We all wanted to get rid of the manual intervention and interaction we had getting applications submitted to either the AppStore or Play Store. What we came up with were a couple of interesting innovations and ways to automate our deployments. First, we configured our source code repository to reject any check-ins without a comment. The Application Lifecycle Management (ALM) consultant inside of me wanted to deny any check-in without a reference to a Work Item, but the developers didn't want yet another task added to their already loaded iterations. Then we restricted the master branch so that any merges had to be from a development branch. All changes needed to be integrated with a pull request and two human approvals. This gave us the ability to inject a code review process into our development life cycle. It was a chance to learn and work together as a team to define our coding standards and to learn tips and tricks from each other.

Once the processes were established and defined, we decided to take on something that we were not sure would be successful; we were going to integrate a security control into the process as a *third* pull request approver.

This was the Genesis of DevSecOps for me.

The Genesis of DevSecOps

Our corporate office had already chosen a SAST tool to use for mobile development so we decided to leverage it for our automation project as well. I installed it into our environment, tested it out with a few source repositories, and I was shocked at the initial results. The majority of our code had major vulnerabilities in it. It was like turning on a light and seeing a mouse running across your kitchen floor. It shouldn't have surprised us though. We had just uncovered what we always felt was in our code by introducing a new level of transparency. It revealed what our true security posture was.

This initial scan became what we would eventually call our *baseline scan* and it captured how many high, medium, and low vulnerabilities an application had. We agreed that these results would define the maximum number of vulnerabilities we should ever have in a scanned applications codebase at any future point in time. Using these values, we created thresholds and configured our build tool to fail a pull request if any new vulnerabilities were introduced by a code check-in.

This meant that if there were a hundred high vulnerabilities, a developer couldn't check-in vulnerability one hundred and one. Conversely, if a developer remediated an issue, the threshold was automatically lowered so no committer could introduce another vulnerability into the system. When we added an Open Source Software Management tool that scanned third-party components into the mix, we eventually brought the code developed in-house down to *defect zero*.

We had now successfully integrated a scanning tool into our CI environment and triggered a vulnerability scan every time we checked in our code. This was a huge accomplishment for the team. We succeeded by breaking down our security silo and collaborating with the DevOps team. What could possibly go wrong?

It was around this time that I made my return to the field of security. I transferred to the security team as a Security Architect, where I would help define tools and techniques required to integrate additional security controls into our DevOps pipelines. With our previous development efforts we had SAST taken care of, it was time to dig into containerization security. We began to make progress before everything changed.

Automated Security

A corporate reorganization rolled our security team into the larger security organization in our enterprise. We had been successful at an extremely small scale but now it was time to take our success to the next level. We would look back later and realize we did nothing extraordinary but just install a tool and hit the scan button with our previous integrations. Now it was time to bring automated security to the thousands of applications we developed worldwide. It seemed like a monumental task but we approached it head-on.

Our initial plan was to implement four security tools inside our DevOps pipelines. The first was SAST, the second would be Open Source Software Management (OSSM), the third - Dynamic Analysis and Security Testing (DAST), and the fourth would be Container Vulnerability Analysis (CVA). I liked to compare these security testing processes to the Four Horsemen of the Apocalypse - Conquest, War, Famine, and Plague. Sometimes it felt like all of these were what we set loose on our DevOps teams.

Before we rolled out our SAST tool, we coded and deployed an area in our cloud provider to host all of our security tools. We then connected this infrastructure to our internal ecosystem so we could provide security tooling to everyone in both our traditional and cloud environments. Finally, we installed our SAST tool and begin to test it out with code developed internally by the security team.

There were many awkward conversations with the teams we were engaging with during our rollout pilot. I believe a lot of this had to do with the belief that traditional security teams were not very technical; but here we were building infrastructure and configuring software. Ultimately some looked at the toolset we were proposing and didn't take it very seriously. It seemed like just another mandate from above in their eyes, and the product owners weren't very accepting either. They couldn't be very flexible in allocating precious development time to remediate security issues. We realized we needed to be cautious while introducing security tooling to DevOps teams.

A traditional development team was the first to use the software. They had a small number of projects that we believed could be set up and scanned in the tool very easily. As scan results began to come in, we noticed that the larger legacy code bases were taking up to forty-eight hours to complete while the smaller applications scanned very quickly. This finding was extremely troubling. Long-running applications clogged the scanning queues and projects were waiting for hours and sometimes days before they were scanned. This alone introduced major security tool drag into the pipeline and blocked other applications from getting to production quickly.

The increased time in the release pipeline was unacceptable to the developers and project managers and caused them to be hesitant in adopting the new DevSecOps process flow.

The idea we came up with to address these longer scan durations was to pull source code on a nightly basis for scanning. This was an interim solution and not even close to our goal of performing vulnerability testing in the pipeline upon code check-in. Regardless, this nightly pull method provided valuable vulnerability analysis and proved that our new scanning cluster was operational.

After onboarding the initial team, we brought a number of other teams onto the system. The biggest issue was that unless we engaged a team directly, they would never voluntarily get on board. We were stuck at about nineteen projects being scanned nightly and it stayed that way until we introduced a program that was aimed at significantly reducing our defect density - the number of high vulnerabilities per ten thousand lines of code in the application.

Within three months of the introduction of the defect density reduction program, we had more than five hundred applications onboarded. They weren't the same kind of large nightly scans that were pulled by our SAST scanner for our traditional teams, but applications that had scanning directly integrated into the development pipeline. This was the start of a massive adoption of SAST by many of our internal application teams. With the collaboration, culture, tooling, and processes we were introducing, we were transforming the enterprise. The seeds of DevSecOps had taken root.

It was an exciting time for the security team when we finally started getting traction with our tools. The problem with this success was that we hadn't anticipated growing so fast - and that's when we almost lost the confidence of all our developers.

Everything ground to a halt.

The Problems with Success

Missing Strategy

One of the first issues that we had to deal with was when we implemented the nightly scans for the traditional development teams we onboarded. We neglected to come up with a strategy to deal with discovered vulnerabilities and false positives. Essentially, we set up these teams with SAST, pressed the scan button, and left them to deal with the aftermath. A similar scenario to what we had left the developers with on our old team.

We scrambled to help weed out the false positives identified by the tool but the scans were still taking quite a long time to complete. Even worse, we were getting support calls on a daily basis from teams telling us that the reports were not working and were hanging when requested. Eventually, we identified that the database of the SAST tool needed to be rebooted on a regular basis in order to clear out temporary files. While not ideal, this solution kept the system operational, and the reports running.

Security Scanning

As new teams were onboarded it quickly came to our attention that they had to wait an unacceptable amount of time for a security scan to complete. In fact, some scans never completed and were caught in our scanning queue where they held up the execution of other scans. Teams began to complain that they couldn't deploy code to production quickly because our security tools were introducing too much drag on the process. We needed to come up with a solution quickly because teams started to remove our security controls from their automated pipelines in order to get their software out the door.

Scaling and Timing

We also had an enormous scaling and timing problem. The scan engines we had configured were not enough to churn through all of the applications that were coming on board. Our ultimate goal was to bring over one thousand three hundred applications onto our platform before the end of the year and reduce our overall defect density. With the problems we were experiencing, it didn't appear like this would happen. We needed a SAST tool, but we needed it to complete scans as fast as possible *and* have 99.999% availability.

More scanners were added and were configured to process one application at a time based on the recommendations of the vendor. Although the solution fixed the queue availability issue, it became quite expensive for the security team to manage the infrastructure supporting the tool. Increasing the scanner count tripled our infrastructure spend.

It became apparent to me that we didn't have a horsepower issue - we had a software issue with the SAST tool. It couldn't support the massive amount of code we were throwing at it, the frequency that we were scanning the code, and it didn't live up to the expectations set when we purchased it. Regardless, after adding extra horsepower we temporarily eliminated wait time in the scanning queues, but we still had long scan durations that needed to be addressed.

Unnecessary Scanning

While trying to understand why scans were taking so long we decided to take a deeper look at the source code to determine what was happening. What we uncovered was that our DevOps teams were not only scanning the code they were building themselves, but were also scanning all of the open source software components that their application required. This was what our OSSM solution was meant to address. As third-party open source components go, many of them have quite a few vulnerabilities and some even critical. Scanning these unnecessary libraries resulted in higher defect densities and the additional volume of source code was responsible for clogging our engines.

The Outcome

By not planning the rollout properly, we ended up spending way more time than necessary chasing down issues with the tool. In fact, the lack of scalability and the expense of the underlying infrastructure to operate the tool hinted that we may have chosen the wrong product for the job in the first place. We had neglected to plan for scaling and had not informed our delivery teams that they shouldn't scan open source software components, Unfortunately, we didn't create a course of action to deal with the many false positives that were being detected by the tool either.

We had gotten the *Culture* and the *Technique* right but missed the mark with the *Tools*. We weren't holding true to the tenants of DevSecOps. With this failure, we could only be partially successful.

Lessons Learned

Have a Plan. Period.

Although we had been successful at rolling out a toolset and integrating it with both DevOps pipelines and traditional development practices, we hadn't considered how to handle scalability issues or the rapid onboarding of so many applications. We simply overlooked the importance of having a plan in place. It didn't have to be perfect, all it needed to be was something we could iterate on as we learned from our successes and mistakes. Automation should have been carefully architected and socialized before any implementation took place.

I proposed a somewhat controversial concept when the greater organization discussed a plan to roll out DevSecOps to the enterprise. I suggested that we may not want to fail a build just because a security vulnerability was found. This may not seem like a very "DevOpsy" thing to do as the phrase "fail fast, fail often" can be found in most books and presentations about DevOps. Think about this rationale though - a security vulnerability doesn't *break* a build. The product is still functional. Failing in this scenario usually disconnects a pipeline before any automated unit and regression testing happens, and ultimately stops a promotion to a staging or pre-prod environment. Why this matters is that as DevSecOps teams we need to take into consideration the operation of the business and efficiency of our pipelines. It's perfectly acceptable to get business sign off on the functionality of a feature while a security issue is being remediated.

To "fail intelligently", I suggest security scanning tools simply tag the build with an annotation indicating that it can never be released to production while vulnerabilities exist. In this scenario, a message should be returned to the developer with the warning, and a work item created for the developer who checked in the code to remediate.

Even though this approach hasn't been widely accepted, it illustrates that security teams need to step up with solutions to performance issues, and ideas to increase flow and eliminate waste. It's just one of the many ways a security organization can improve their "street cred" with DevOps teams.

It's also just as important to advise DevOps teams to avoid immediately failing a build based on the output of *any* security scanning tool until false positives are identified and removed from the results. If this isn't done, you'll get a large freak out email from the product owner asking how the team can possibly remediate 5000+ vulnerabilities without compromising their deadlines.

Finally, we should have allocated time to assist teams as they started to use our tools. Our plan didn't account for any training that may have helped teams better understand the output of the scanners. Instructor-led or computer-based training could have helped teams with adoption and minimized their frustration.

Know where you are and where you should be

We also learned that we needed to implement a timer around every scan to determine how fast it would run from start to finish. This measurement evolved into a Key Performance Indicator (KPI) that allowed us to measure the speed of our scans, identify if the measure was outside of an identified threshold, and alert the appropriate teams of any issue.

An acceptable execution time was determined to be less than five minutes for a full code scan, whereas an incremental scan that only detects vulnerabilities in a changeset should take somewhere between thirty seconds and two minutes to execute. If these thresholds were out of range, we would do a deep dive into the code base and determine if we were scanning open source components or assets that didn't need to be scanned. We also suggested to the DevOps teams to break up a project into multiple smaller scans to increase flow. If an application wasn't architected as a microservice, the team could split their code base into multiple scans by excluding specific directories.

Unfortunately we also forgot to implement availability checks to ensure that our tooling was no less reliable than the build server; something that a Site Reliability Engineer (SRE) could have helped with. It's extremely important to have the proper instrumentation in place to proactively monitor the health of an infrastructure. This ensures that the right individuals are informed and the appropriate actions are taken to keep the system healthy.

A fool with a tool may still be a fool

SAST tools have the capability to scan open source software for vulnerabilities, however, you'll find that choosing the correct application for the job will reduce scan time and the overall effectiveness of your security solutions.

Leave open source scanning for the tools that are available to do it in an appropriate way. This was an extremely important lesson we had learned. OSSM tools don't need to walk the codebase from the point where an exploit can be executed to the function call where the vulnerability was introduced. They effectively look at the signatures of the third-party components in use to determine which components are vulnerable and which are not. This is what makes these tools so effective and fast. Some of these toolsets even display the history of a component and the upgrade path a developer needs to take to remediate discovered vulnerabilities. They can even automatically remediate the issue for you if binary compatibility can be maintained.

Finally, even though we engaged the vendor to help diagnose the issues with stalled report generation, we never found out why the SAST database needed to be restarted. They had no explanation or solution to remediate the problem. We tried increasing the computing power and capacity of the database but the issue still remained. We eventually created a Lambda function that would execute at regular intervals to cycle our database instance. It was a tiny and temporary band-aid for the larger product defect but it worked. The lesson here is that purchased software most likely contains bugs that you'll need to design a creative solution for in order to circumvent.

Don't Fear the Four Horsemen

Implementing a successful DevSecOps program is hard. Really, really hard. Accept the fact that you're going to fail, your solutions will fail, and that there's a balance that needs to be maintained between availability and performance. Learn from these failures and don't fear The Four Horsemen of the "*DevSecOpalpse*". Begin your evolution to a DevSecOps culture by integrating security into your DevOps pipelines one tool at a time. Integrate each as best as you can, and move on to the next control while you iterate.

Even though there are risks integrating any tool into your automated pipelines, there's nothing that can't be overcome with a bit of planning and patience. There's a choice to make when addressing acceptable risk. Do you accept a 10% failure rate when your team puts in 100% effort, or do you put in 0% effort and accept 0% success? Risk is always present but without quantifying and accepting it, you'll never have the chance of being successful. Take the plunge, face the challenges, and learn from failure.

Finally, ensure a plan is in place that addresses what could happen when moving towards DevSecOps. We began by scanning a handful of applications and then expanded to over five hundred overnight. We weren't prepared for the load. How do you plan to support over a thousand applications? What about two thousand?

Final Thoughts

We really hadn't anticipated the rapid adoption of the security tools we were standing up for our DevOps teams and this oversight created issues the security team scrambled to address. Even though our culture had evolved over time and we iterated to perfect the technique, we ultimately realized that the toolset we selected for SAST was less than optimal.

At the end of the day, our entire Enterprise was on board with integrating the tools we were providing. We built it. Everyone came, and we weren't ready - This turned out to be our problem with success.

Acknowledgements

I have to start with thanking my wife Nikki. The endless support you've given me over the years enabled me to explore any and all opportunities where I could learn and innovate. To my children, thank you for pulling me away from the computer to experience the beauty and wonder of life and fatherhood.

Thank you to all of the proofreaders that helped create this chapter; especially Ivan A De Los Santos, Predrag Petrovic, Graeme Park, and Adrian Lane. Without your help this chapter would have read like an alphabet without vowels. I'd also like to give a special shout out to Stefan Streichsbier and Edwin Kwan for coming up with the idea to write this book in Singapore while the street food I was eating coated my mouth with dragon fire.

To Michael Trofi, thank for proofreading, but more importantly thank you for the many years of friendship and security mentorship. You alone are responsible for convincing me to go legit and use my hacking skills to help secure the world, and for that I am truly grateful.

Finally, a heartfelt thank you to Mark Miller. You gave me a microphone and I found a voice. You've shown me how to observe cityscapes in a reflective way. Thank you for challenging me to improve, and for being the Tribal Leader of our DevSecOps Family. You are a true friend, an amazing storyteller, and probably the most interesting person I've ever had the pleasure of meeting.

About DJ Schleen

DJ is a DevSecOps pioneer and currently works as a DevSecOps Evangelist and Security Architect at a large healthcare organization. He provides DevSecOps thought leadership throughout their journey of cultural revolution and digital transformation. DJ specializes in automating security controls in DevSecOps environments and is an ethical hacker as well – doing significant R&D work in Moving Target Defense, Mobile Security, System Exploitation, and Penetration Testing.

As an expert in Application Lifecycle Management (ALM) and the IT Infrastructure Library (ITIL), DJ has worked to streamline development pipelines for many Fortune 100 organizations by focusing on people, process, and the right technology. He is an active speaker, blogger, instructor and author in the growing DevSecOps community where he encourages organizations to deeply integrate a culture of security into their core values and product development journey.

Chapter 4

The Tale of the Burning Programme

by Aubrey Stearn

Chapter 4
The Tale of the Burning Programme

Unfortunately dear readers, this journey isn't quite akin to young Bilbo setting off in *The Hobbit* for his epic adventure. No! We join our heroine slightly later in her quest, Midgard....only kidding....mid programme, I'm going to stop with these mythology references because I only have about 4 left.

Our heroine has joined a programme, a programme that has been running for 18 long months. Adverse to in-house development, the somewhat common, traditional view has been taken that using partners or system integrators to deliver programme components is akin to de-risking.

As an archaeologist must work backwards from the present to the past, we will be deconstructing the scene to uncover all the skeletons, review and analyze the evidence, and formulate a hypothesis before we're able to work forward and start delivering value.

We start at a time in my teens when I was working for a karaoke club that was running upon a FoxPro database using an unstable Delphi client.

This system was absolutely loaded with weird hacks to handle the multiple languages both on the client and in the database. In rebuilding this system, perhaps about the 3rd day in, it still wasn't finished and I remember walking home feeling disheartened. I received a telephone call from the owner of the karaoke club and I calmly explained why it was taking so long to resolve the issues due to the various unicode hacks involved. The guy was extremely upset at me and did not accept or want my explanation. My heart sunk and, being so young, I was unable to respond in the concise and professional manner as I would today.

I was dejected for failing him and letting him down. On the verge of tears, I called my dad and explained the situation. He said to me, "You're telling him a lot of story, and in life when people pay you to do something, you will find they don't give a fuck about the story when you don't deliver on time."

My dad made a good point and it is a life lesson that has remained with me to this day. Everyone has this same story and on a really long-running programme prone to failing and crashing, , you can be certain the wisdom of my father applies in triplicate.

I usually find there is at least some grain of truth to all stories. It's a good idea to hear everyone's side during discovery (or at least as many as you possibly can) until they decide to stop talking.

Our biggest problem is DevOps

Back to our programme, I remember meeting the integrators development manager for the first time. I asked him "What do you think your top problem is in development?", he replied, "DevOps is our biggest problem".

Every time I hear this (and it's way more common than you might think) I know immediately that the problem is development, I've yet to be wrong on this assessment.

While reviewing their code repositories for the first time, I uncovered hundreds of long-lived feature branches in some repositories and, "branches-as-folders". This could be a new thing!

I had a meeting with the 'DevOps guy' and asked my usual questions;

- How many pipelines do you have?
- How do they work?
- What are they running upon?
- How is their continuous integration (CI) and continuous development (CD) structured?

Things started to fall apart here, this guy had apparently only 10 minutes to talk to me. The System Integration manager had told him so. He wanted out of there as soon as possible.I called the programme leadership mid-conversation, who called the SI leadership, and quickly booked him out for another 45 minutes. This was a mild power play on my part.

I'm not going to lie. I was pretty annoyed by his 'I've only got 10 minutes for you' attitude, but I kept my cool. It's not the sort of attitude I appreciate.

I already knew the programme wasin trouble. The faster I can figure out why, the faster we can start making it better which is good for everyone... DevOps Guy included.

Take-homes from this conversation:

- 1 Jenkins instance per environment
- Octopus also in the mix per environment - NUGET not involved at all?
- No idea what version of Jenkins they are running
- Devs don't know how to use Jenkins
- No CI builds for anything!!!
- No sign of Blue Ocean
- Deployment of code to an environment is entirely manual
- All environmental secrets and credentials for everything are stored in source control, they have never been rotated, almost 100% of the staff on the SI side have been rotated.

Critical Thoughts:

- If we're an Azure/Microsoft house why on earth are we building self-hosted Linux boxes running Jenkins that need to be maintained and hardened.
- Why are those self-hosted boxes publicly exposed over port 80?
- Why do we have both Octopus and Jenkins?

- I walked through a deployment yesterday that failed because of a mistake in syntax that was sitting in the branch for 3 weeks...why are we not doing CI builds?
- Why doesn't the "DevOps Guy' know what version of Jenkins is running and if they are using pipelines.
- Essentially VSTS has way better alignment with the technology set, deep integration with Azure and KeyVault, will happily pivot between *nix and Windows technologies.
- Finally, Release Management will make deployments repeatable and shift control from development back into the product domain.

Discover like a Mother

As is usual with these sort of firefighting exercises, I spent my first three weeks doing discovery. Longer is always better when it comes to discovery. However, we're also against the clock and need to show some value or insight.

Now, in the discovery phase for a typical enterprise programme, I'm looking for some core artefacts or statements:

- What is the Deliverable or Objective of the programme
- What is the Deliverable from the System Integrator (SI)
- What is the working relationship between the SI and the Client
- If this is Enterprise Architecture led then where is the High Level Design Document(HLD)
- Who is responsible for what

If you don't know me, I'm the sort of character that usually has a Head of DevOps title but I generally sit over software development as well. After all, writing high-quality software is imperative to making your DevSecOps movement a success.

I'm already sensing that this programme was borne of architects somewhere so there is more than likely a High-Level Design document knocking around. The first job is to get hold of the latest version of that and walk it.

This is the first tell-tale sign something is wrong: no one knows where the latest HLD actually is. There is no central document repository and it turns out the SI is actually using Confluence to store virtually everything, but as the client, we don't have access.

Worse, upon reviewing the contract, we aren't even entitled access to their documentation. This is a major fail, increasing the amount of time I need to do discovery and increasing the amount of hearsay introduced.

I'm currently sensing notes of bad communication, poor contract writing and lack of 'technical oversight. My father would say in the style of Jilly Goolden and Oz Clarke: 'a bouquet of boot polish, used nappies and raisins' which I think is just his way of saying something doesn't smell right.

A symphony of fuckups

Three days and a couple of emails to programme leads later I sense some movement as in my inbox appears a copy of the HLD that I can finally read through. Boy is it a masterpiece. I want to call this "A Perfect Symphony of Fuckups".

The first obvious thing here is the design includes a dual data centre with failover, nothing is cloud native. I'm not seeing availability zones or components spanning regions, I'm seeing databases that need to be switched over to become active.

Enterprise engineering has decided on a service bus architecture, and I find myself thinking, "Dear Lord this will be expensive" before it's even used. Now don't get me wrong. There's nothing wrong with a service bus architecture; however I do advocate the principle of engineering for today's problem and not tomorrow's extensibility or modularity.

The number one problem I could see with this design: we're 18 months in, there are 480 somethings hanging off that service bus in 8 different environments, costing 30k British Pounds Sterling a month! Given the programme at this point is overrunning, we will easily hit 24 months with semantically one application on the service bus. That's another £180k squandered for extensibility.

We could go on and on about this but it would make extensibility and modularity sound even more stupid, overhyped and unsuitable.

The difference between inter-API calls with a couple of well-placed queues and peeking at a topic in code is literally a few lines. In terms of programmer time that's going to be significantly less than the cost of going service bus.

Moving on to the great business rule of fact, fallacy or fantasy', is it realistic to change a business rule and not affect downstream components or development? From my experience, most organisations would benefit from a quality in-house development capability who can just update the business rules as code when desired.

Forms Builder! Another strange component in this stack. Looking through the requirements I can't honestly see a driver for this other than they have lots of forms. No requirement to store data submitted short of it ending up on a queue or dead letter queue on failure. I did a little experiment with this component, switching one of the forms to just a plain old angular form backed by a service. Turns out actually it was faster to build the form with angular and it had the benefits of being testable. Once again, there was no testing strategy for the component which had an overall impact to the testability of the front-end application.

Building on sand

At some point, you have to wonder how a programme of this size with an objective that is basically a C.R.U.D operation over a CRM has got in such a convoluted state. Forensic analysis of the of the various repositories revealed many truths, the most damning of all: the 'Demo' repo!

You know what this looked like? Two architects went into a room and dreamed up this ideal architecture in an afternoon and got some people to knock it together. What annoys me the most about this is that pretty much all of the core technologies chosen for the solution are features in this 'Demo' and you know what's missing and stubbed out? Testing...Doh!.

I'm sorry to sound like a squeaky wheel on this, but can you imagine the legacy that would have been prevented if they had just tried to actually test the technology they were suggesting. I'll say it again, architects need to touch the metal. Almost nothing fits into pretty little boxes.

What was interesting was the lack of development standards, lack of implementations to control standards, lack of definition in the contract to hold the supplier to those standards, essentially the foundations of this programme were built on sand. For all the frameworks that exist on this programme, it surprises me the most that one proven for successful development is missing. I suspect this programme has a shortage of modern agile development experience.

The Netflix dilemma

I read a lot. I frequently use the things that I read to form the basis of a hypothesis which, in turn, will become an experiment. If you're an avid reader, especially on the technology scene, you might have read a book called "Powerful" by Patty McCord. While there are a lot of things in Powerful that resonated with me, there were equally a few things that didn't.

There have been a few times in my career where I've said that while someone was the right person at a particular point in time, they are no longer right for now or where we are going. This is a harsh truth of leadership: not everyone who starts the journey with you will be there at the end.

Sometimes people are super happy doing as little work as possible and when you ramp up the pace, they get very unhappy that the ground has changed beneath them. Some people like to be just a number cranking out a couple of Jira tickets. I knew a guy like that, being in a high performance team just wasn't his thing.

The 'everything is shit guy'(or girl), do you know them? Have you been them? I certainly have, borne of lack of action when I said something was wrong and proposed a meaningful solution. It's a cardinal error for you to know something is wrong but do nothing about it. Honestly, every time I've heard that line it drives me crazy.

Hell, last night I was talking to my girlfriend before bed and she was telling me about a dressing down she got from her manager. She works in a retail role, she's the assistant manager and, like a lot of retail positions, it's at a small concession. They have a high turnover of staff. This can be attributed to University and a preference for part-time workers.

My girlfriend told me she got a dressing down for charging the price on an invoice for a battery replacement rather than the price plus postage.

I said to her, "Why don't you email head office and tell them to add the postage price to the invoice so people won't forget or have to remember this stuff." Given the volume of staff turn over, I had assumed this would happen quite often and this process knowledge can get lost over time or, in this case, simply forgotten.

She said to me, "But they won't change anything." In itself, this speaks volumes about the company culture. I said to her, "At least if you send them an email and they say no, you've held your self to your own high standards and principles rather than lower yourself to someone else's".

To me this is a core point to call out, it's so easy to be like water and go with the flow but all you're doing is lowering your own standards and taking the easy road. If you give people power, ask them to be agile and they tell you something is broken and you ignore them, they will become disenfranchised. You will birth a 'Everything is shit' person (or worse, team!).

This programme has a lot of people who are past their sell-by-date, lack technical discipline/experience/relevance, it also has many chiefs, hardly any indians and people are used to talking in abstract rather than in actions.

You're a purist

I pretty sure that at this stage of my life I'm a pragmatist. Yet here I am with the programme lead being told I'm a purist for suggesting replacing a technology at this late stage.

Now coming back to this point about testing and good testing, if you've chosen technologies that are difficult to work with, verging on unmaintainable, incredibly difficult to test and violate the do not repeat yourself rule an awful lot, this is one of the few times I will suggest replacing a technology.

Don't use a Barracuda firewall when no one knows how to maintain or set one up, use the cloud native solution; Application Gateway on Azure, Cloud Front + WAF on AWS or maybe a combo of Cloud CDN, Amour and Global Load-Balancer on GCP. Don't pick technologies that sound like the right thing because they were the right fit when used on-premise.

Now don't get me wrong, maybe you have a legit requirement for a Barracuda, but I hope that hasn't come from some ancient security people who just don't get it anymore and should have been recycled with yesterday's milk. You don't get to sit back when you're in security, you don't get an easy ride. Security is a moving target and it never stops.

If I pop my CTO hat on for a moment and think about the sort of engineering profiles I will need to hire, having to get someone with Azure + Barracuda seems stupid if there is an Azure native solution available. I can just hire a standard Azure engineer without increasing the technical footprint of the overall engineering profile as well as the cost. The technology we choose will always have longer term cost ramifications. A great example of that right now is that an AWS Cloud engineer can be found for a lot less than an Azure or GCP engineer.

Seek forgiveness not permission

When I join an organisation, it's with a particular mandate and you can be sure I'll execute on that until I'm told not to. Eventually, on this Titanic of a programme, we reach a point where enough of a picture has been painted that we can understand how to quickly go from A -> B.

I'm being told that things which would normally take me one or two weeks to complete, will take months to achieve or need to go into a release next year! As you can tell I'm not the kind of girl to take bull-shit or ineptitude well, so making good on my mission I will stand up a team with the key skills I need.

"We do this, and then we can do all of these things and now it's all automatic. Quality is mandated, you can't even contribute without maintaining our quality standards." This is the mantra that person will live and die by in your world. Everything is simple, quality starts at the beginning....there are clearly times when I wonder if I'm a cult leader.

Let's talk about my cornerstone, that one person who works super closely with me, that person who learns the narrative, to whom you teach that critical thinking.

In order to build a team I'm going to need them to buy into me as a leader and a capable team mate. One way to do this if you have a existing team is to work with one person on that team and go on the whole journey together, that person becomes the cornerstone, choose wisely, they need to be a team player, choose Mr. Solo and all of that great work will remain in their silo of one.

If I can take one person all the way to that epiphany moment where they can articulate why the architectural design plus development process is wrong and, give them the tools to do it the right way, finally we do it the right way, albeit just a tiny slice.

We constantly reinforce our point, by referring to the work done by our Systems Integrator. It took us 5 minutes to build and ship that docker image and Kubernetes deployment, the SI currently takes 2 days...cue the funny look and "Wait, everything you did was so simple," comments.

Taking control of the narrative

The trick with wrangling back control of the narrative is to tell a compelling story in such a way that if anyone is to merely drift from your story line the very notion of anything else seems stupid, we call this framing.

Apple is very good at framing. By setting the frame, they can control the bounds in which you operate. We might have an S year when the only thing that really changed is the phone got a better camera, but oh boy can we talk about the camera for an hour and a half.

You probably want to deliver this like an eight count dance routine, click those fingers girl! Make it snappy. If you give people a moment to think, they will. Keep hitting them with progress.

So let's take the story of a nodeJS microservice and it's 8-count:

1. Npm test
2. Pre-Push Hook run Lint & Test
3. Push Branch -> CI
4. Raise PR -> CI -> Merge
5. Release Management deploys into development
6. Kubernetes rolls new container into service
7. Service logs to elastic search
8. ChatOps bot pipes into slack channel

How long did that take? Minutes. When I show you this I'm going to carefully narrate the experience. I know how long those docker image builds take, especially on a freshly built host with no cache and I have an equally long explanation to fill your mind with my narrative.

Before you know it, we will have taken a new line of code through its journey from local machine, to repo, to pipeline, to release management, into its new home in development and confirmed its observability.

Now, with your head filled with my story bound within the frame we defined, I'll draw that quick comparison to the 3-month deployment, even the 3-day deployment, versus what we just built and, to boot, we built that from scratch in 2 weeks, while teaching a team of people to do the same over and over again. We taught men to fish and caught enough to feed the town, a massive success story that we can tell over and over again as a legend.

This is a campaign of hearts and minds. After 18 months of failure not only have we succeeded, we've invited you to join us and watch. We want to show you what we have built and how we have done it.

I didn't magically fix a whole programme or even every single component, but I did set a clear boundary to tell a specific story, a powerful software development story, one with a beginning and an end.

Don't underestimate the power of this cultural shift. We're inverting everything about how our Systems Integrator is working with us.

The narrative says we now have a internal team that is capable of a quality software development process. That team was able to take a small shim of functionality that was originally written as 15 Logic Apps on a Azure Service Bus, replace them with four well defined microservices with incredibly strong testing and well defined coupling.

Turning a corner

Repeating success brings affirmation that we're doing the right thing. Being able to crank out new microservices in our own development capability was huge.

People started to tell me how they had seen changes in people in my team, that they were smiling and talking about how much they loved what they were doing. Suddenly people are being actively told to come over and look at what these guys had built.

You can't underestimate how big this paradigm shift was for the business, to spend 18 months without playbacks of sprints, or features developed when suddenly this team from nowhere begins asking people to come and look at this story we were now able to tell.

Things that I wish I could fix but couldn't

Product

I wish I could have brought someone phenomenal into to run product. Sarah Longhurst & Colin Houlihan are two of very best product people that I have ever worked with. I've always seen them as product first and domain knowledge second. Both have taught me vast amounts about delivering tech as a product. One unyielding point from Colin was about what moves the needle for the business, the metrics that matter, how we measure and track them, using them to inform and set the tone/direction of development.

I won't go into how to run product because it would easily fill a whole book and I'd need a lot of help from people who are much more talented than me. While I run my team like a product-centric team, I'm sure there is a lot that I still need to learn about product.

One of the core failings of this programme was that the product was so distant. At some point, core deliverables had been defined early in the process. I'm not sure how you can really know all of those up front. These were enshrined in a contract with the supplier. Then the product people would leave the scene and pick up again during testing...w00t!

Product is the nucleus and development are the electrons. Product guides us every day. We should all be watching the metrics underpinned by our product. These are normally business aligned, transactions, conversion etc. If you're the platform team, maybe you're watching feature request volume, average build queue depth/wait time and platform availability.

Development is Dev, Sec, Test, Ops

My major frustration with this programme has been the language used to define development. Development means more than writing code. It means ownership of testing, security and the deployment work. When I first heard that it took 3 months to deploy, that in itself told me a lot about the mentality of the development team.

During discovery, I'd talked to the majority of developers or at least the people we had access to in the UK. When I asked them if they knew how to test or if they were experienced in testing, the answer was a resounding "yes". My follow up bonus question is as predictable as a bull charging a matador, "Why aren't you testing then?". I won't dignify this chapter with their answer. Needless to say, it wasn't acceptable.

As a developer, you have a personal responsibility to practice your craft properly and at a high level. That means you own testing. Absolutely nothing should be getting pushed to master without 100% coverage and an adequate amount of negative testing.

But Wait! There's More!

Wait! Don't stop there. In fact, I hope you thought about this first: attack vectors. Test for those too. Is your server spewing out "X-Powered-By" headers? Are you validating schema payloads? Are you whitelisting non-alphanumeric characters? Know and understand your security vectors and test against them, both unit and integration as well as external.

I cannot overstate my level of frustration with the head of development everytime I would demonstrate what good testing would look like for each component only to hear him wax it up as version 1.2 or 1.3 'nice to have'.

Let me tell you something about not testing.

Not testing works great right up until you have to change one thing. Then you'll KNOW what creek you're floating down without a paddle! When you're not testing, you're constantly moving forward which is why it works. You never have to go back and make sure everything still works.

When you start to change anything the untested house of cards begins to fall. Personally, I always find good testing also helps with DRY (Do not Repeat Yourself) violations; if you find yourself copying and pasting tests, that's a good argument to move that code and the tests to somewhere shared.

Culture

The lack of a transparency was absolutely brutal. With development locked away in two rooms with the doors always closed, could you imagine an environment less conducive to engagement? There were no flow walls anywhere, and no Kanban boards anywhere. During discovery no one knew what development was actually working on. Everything I hold dear to good development was thoroughly broken.

Being contractually restrained when it comes to fixing development was one of the most painful, and sometimes emotionally distressing, times I've had as part of my work for a long time. It's given me a couple more questions to ask when I interview with prospective clients.

I am happy with the culture that we stood up within my team. I can't tell you how many times I said 'I've never done that' or 'I don't know. It's really important that you have a welcoming attitude that makes things like that OK to say. Development and maybe other areas of technology have a long history of learned behaviors that need to be unpicked. Behaviors that don't foster a nurturing environment, don't promote pairing. Behaviors that are simply unhealthy for good software development and promote a solo player culture.

A quick mention on the subject of RockStar developers; there are a lot of mixed feelings towards the subject. I guess it depends on your definition of RockStar. It would seem to me that a lot of people use it in the negative, a toxic, self centered, solo developer, that doesn't play well with the team and is a bit of a diva.

I've met some genuine Rock Stars in my career. Leonardo Fernandez Sanchez is one of them. Leo not only had the infinite patience for me when I was nowhere near as good developer as him, but he also put up with my tears in the first year of hormones.

I've honestly never felt more stupid than having to learn a new testing framework on the fly with someone who knows it so damn well. Leo reminded me it took a long time for him to become an expert or even look like one. Testing isn't a skill we practice consistently, we become great coders by doing it all the time. I know Leo would remind me to tell you dear reader that practice is what we all need, to achieve mastery.

I think that was personally a very hard year for me. The first year of living full time as a woman, first year of hormones, not only was I lacking confidence as a woman but I'd lost confidence in myself technically more than a few times.

I remember a great point Leo made about squash vs not squashing. Take into account your rollback strategy. One commit will be a lot easier to unpick than multiple commits.

My favourite Leo quote is 'We optimise for deletion', a line never more welcome than when I'd spent 2 days listening to someone berate my code within earshot. I ended up crying uncontrollably and felt so hurt personally, all I could think about were the evenings I'd put into that feature because I cared and enjoyed the problem I was solving. He reminded me that while yes slagging of someone's code when they can hear is bad, accepting that when someone improves on, or replace/delete code is a good thing. Accepting that means you're out for the team and not yourself.

C-Level Engagement

For all the CIO, CEO, CFO, CxO's out there, I wish I could help you to see the love we have for this programme, all the people that want to help realise the original objective. Alas, you are protected by so many levels of self-serving executives I couldn't find the right vector to engage you. I wanted you to know what a massive impact it would have on my team if you came and walked the board with us.

If the programme exists to help the business achieve the objective you guys lay out, then isn't it a powerful message when you come and find out from the people at the coalface how your programme is performing? Come and tell us how important that objective is to your strategic objectives.

Equally, if you ask us how things are going, that sends a powerful cultural message. "Communications are open!" I'm knocking at your door saying "Hey buddy, what's the news, how you doing?"

I think it's always interesting when engaging with your execs. Your success can depend on a number of factors. A technology transformation can and will touch the entire business which means a huge part of transformation is the hearts and minds piece.

If you're being directly engaged by C-Level execs, then you might have that explicit trust already. With an established track record, the buy in at top level is already there and it's about everyone else below C-Level.

If you're being engaged by someone below C-Level, then we might need to do a couple of greenhouse size experiments in the middle of the business. We can get a few wins under our belt, increase the appetite for a bigger risk profile and get those C-Level execs onboard for the rest of the journey.

Either way, executive sponsorship is crucial to the success of any transformation programme.

Recognition

The great work my guys have done needs much recognition, not only because I know how hard it is to measure up to my standards (and how hard they try to meet them), but because they have risen to the occasion every time, even in the face of insurmountable self-doubt.

I am proud, like a mother. Proud of my work family and I need to reward them. This is maybe less aligned with what I've read about Netflix culture and more aligned with Google and Facebook, but I just do things a different way.

Sometimes it's not a huge thing that you need to do to show thanks. We did so much work on this project with Windows Subsystem for Linux and while that was great and an interesting experience, working through the pain I wasn't feeling on MacOS was a little unfair.

How much would it cost the organisation to get them a MacBook? Not much. Sticker that bad boy up. They know the development flow at this point, and work predominantly in the CLI anyway. It's not a traditional gift, but honestly, that would have made their month! Not only does the team feel great and get recognition from the business, but we potentially save a ton of man days of churn.

Recognition is a two-sided thing. I want the guys to genuinely see the contribution they have made because we are our own biggest critics when it comes to measuring our improvement. Then there's the other side of the coin which is me, your co-creator, your partner in crime. The person telling you that you can do all of these things that you don't believe.

I want....strike that, I need you to understand that this doesn't work when I'm a one girl army. It works when we are a team and I can rely on you. When you bring that to the table it honestly makes my day When I see you happy it makes my day. I want you to enjoy working with me even more because I'm enjoying working with you.

Wrapping up

It's been a crazy journey... so many technical missteps, so many characters. I think if I had to leave you with a couple of core take-homes it would be these.

If you're doing Scrum-But, Wagile, Fragile or some other crap, just stop. Stop the lies, stop the fallacies, and go all the way back to the basics.

Define done! If you have 8 environments, then your definition of "done" is that it's in production. You don't need to turn it on, but it should be shipped and ready to go.

If you have test and QA in the mix then that is included in "done" Operating as a single body is imperative to your success. If part of the process is letting you down and your 1-2 week cycle is taking 3-4 weeks, then you are now all invested in improving the flow or removing those environments or shifting those who lag behind left.

Works-in-Progress limit of one. Use a Kanban board. Take a single feature and run that through as a mob until it's done! You need to understand what your cycle time looks like, and how long does it take you to get a feature through to done.

This exercise is about honesty. How long does it take us to actually do something vs our estimation. How effective are we as a team at building something? Are we working with product owners properly? Does product know how to work with development? You don't have to absolutely follow any methodology, but being able to run Kanban and conclude accurately what your cycle time looks like, is a powerful exercise.

The best thing about agile is how quickly we can manifest value. The smallest exercise can make a difference. If you have zero percent test coverage, then add 1% and you are getting better. It's addictive and you can build on it. That is a change you can make immediately.

The Outcome

So what happened in the end? Well, with all of the fantastic work we did, circulated, popularised, and framed, we had enough collateral & trust with the business to go into the development business internally for ourselves. We hired a few more fantastic people with experience who had an absolute fetish for testing and #TestInProduction.

We ditched the service bus idea until it was needed and just built a set of robust API's with great logging and observability. We phased out the system integrator's efforts and asked them to call our microservices instead.

While there were plenty of things that didn't work, we never saw regression in any of our code. Releases took minutes and we saw as many as 30 per day. We stopped hearing about regression packs and started getting compliments about our robust testing.

Agile is great, follow it and you will produce equity. You can use that for experiments & mistakes or bank it for a later date. It represents the trust the business has placed in you to do well by it.

Be bold. Lead change. Build honesty and transparency into the DNA of your culture.

About Aubrey Stearn

I'm Aubrey. If you're in the DevOps scene in the U.K., you might have seen me about sweating on stage. I love what I do, I love the DevOps movement and more than anything I love my own brand of DevOps which is a completely picture of code to done with dev owning the whole process. I'm also transgender, 6.8ft tall and an ex cage-fighter.

Chapter 5

Threat Modelling – A Disaster

by Edwin Kwan

Chapter 5
Threat Modelling – A Disaster

This is a story about how, regardless of the best intentions of all involved, we got it very wrong before we managed to get it right. It's a story about one aspect of our journey to "shift security to the left".

Background

We started life as a sort of "quasi-bank", a financial start-up focused on merchant card payments processing services for small to medium-sized businesses. We provided payment terminals to our merchants so that they could accept card transactions (such as credit and debit cards) from their customers.

The founders were three engineers and we had a relatively large development team. We had built our own payments platform and the ownership of our own technology was seen as a competitive advantage. From very early on, the engineering team had adopted the XP agile methodology. This encompassed practices including pair programming, test-driven development, and continuous integration.

Pair programming is the practice where two people are working on one computer in constant conversation and collaboration. There are two computer monitors with their displays mirrored, two keyboards and two mice controlling the same computer. Within a team, individual developers 'pair swap' across tasks. Depending on the team, the pairs can swap either every day, or every other day. It's a very social way of doing software development but does not suit all personalities. Pair programming means that your code is constantly being reviewed by your partner and hence should have fewer errors. It's also a great way to share technical and domain knowledge.

(Pair programming in action)

Test Driven Development (TDD) is where the implementation of any feature starts with a failing test. When applied strictly, TDD mandates that no code can be written unless there is a failing test. This may be a small unit test, a larger 'acceptance' test, or anything in between. The important thing is that all of these tests are automated and run on every build. This gives us confidence that any code changes have not broken any existing functionality.

We also practice continuous integration (CI). This is where instead of waiting until the entire feature is developed before pushing the code to the master repository, we favour small incremental code changes. And because we do continuous integration, we also have a build pipeline. That means whenever code gets pushed to the main repository, the build pipeline is triggered. The build pipeline makes sure that the application code compiles and runs all the necessary tests, making sure that everything works and that the application is backwards compatible. If everything passes, a new release version for the application is created.

As well as following the XP agile methodology, we also didn't believe much in documentation (DRY), preferring the code itself being the source of truth, and preferring face-to-face communication over documentation. We felt that documentation was unnecessary as it was time consuming to create and to maintain. It too often becomes outdated and incorrect not long after its creation. We believed that if you wanted to find out about something, you could either speak to someone who had the necessary knowledge or look at the code to get the answers. This approach encourages the team to write good quality code that is simple and easily understood. It fosters communication between teams and allows them to move fast and easily adapt the architecture to new requirements.

As for processes, we believed that if you want a particular process to be followed, it needs to be the only possible path that can be taken. Let me share an example of this. We wanted all code changes to be performed against an existing task. Since we use Atlassian's JIRA for tracking all our work, we wanted all code changes to be made against a valid and open JIRA task or issue. We use Git as our source code version control system and one of the many features of that system is that it allows checks to be performed when changes are made. Those checks are called Git Hooks and we created one to enforce the requirement that all code changes need to be made against an open JIRA task. The Git Hook we created was a commit hook and it validates that a JIRA ID is included at the start of the commit message when any code changes are committed. It also checks the status of the JIRA task, making sure that it is in a state that allows work to be performed against it. As such, there is only one possible way code changes can be made. Without an open ticket in JIRA, Git will refuse to commit your changes. Another example is the requirement that newly released versions of an application are backwards compatible with the version in production. This requirement is enforced post-commit by a suite of tests that run in the build pipeline and fails if the release is not backwards compatible with existing API contracts and database schema.

Why we introduced Threat Modelling

It all began a few years ago when we embarked on our security journey in software development. Why did we embark on this journey? Because of changes in the regulatory environment, we decided to try to obtain a "full" banking licence. Becoming a "full" bank would mean that we would be able to provide banking services: providing transaction accounts and offering loans to our customers.

Providing transaction accounts to our merchants would mean that rather than having to wait a few days while funds from their card transactions were transferred to an account with another financial institution, they would be instantly available. And being able to offer loans would allow us to help our merchants grow their business. One of the pain points our merchants shared was, being a small to medium-sized business, it was very difficult for them to get a loan.

However, there is an increased level of risk associated with becoming a "full" bank, so we had to evolve the company and our practices, including our risk management and application security capabilities, to be able to handle it. This was especially important because we had made the decision to build our own banking platform.

Our payments platform, which had been built a decade earlier, was composed of several large applications. That was how you built applications at the time. The state-of-the-art in technology had changed since then. At the point that we started building the core banking platform, the industry had moved on to micro-services, so we decided to design and build it using a micro-service architecture.

Using this style of application architecture means that instead of creating a huge monolithic application that does everything, we create many smaller, single function applications that work closely together.

(Example of a microservices architecture diagram. Note: The connectivity between services in this diagram was done by a colleague's daughter, who scribbled/drew the lines between the boxes.)

The first thing we did was to get some consultants in to obtain a baseline security assessment of our SDLC (Software Development Life Cycle), or as we would later refer to as our "SSDLC" (Secure SDLC). For the assessment, we decided to benchmark ourselves against the guidelines for "secure software development" set by the country's financial regulators. One of those guidelines said that we needed to consider security at every stage of our SDLC.

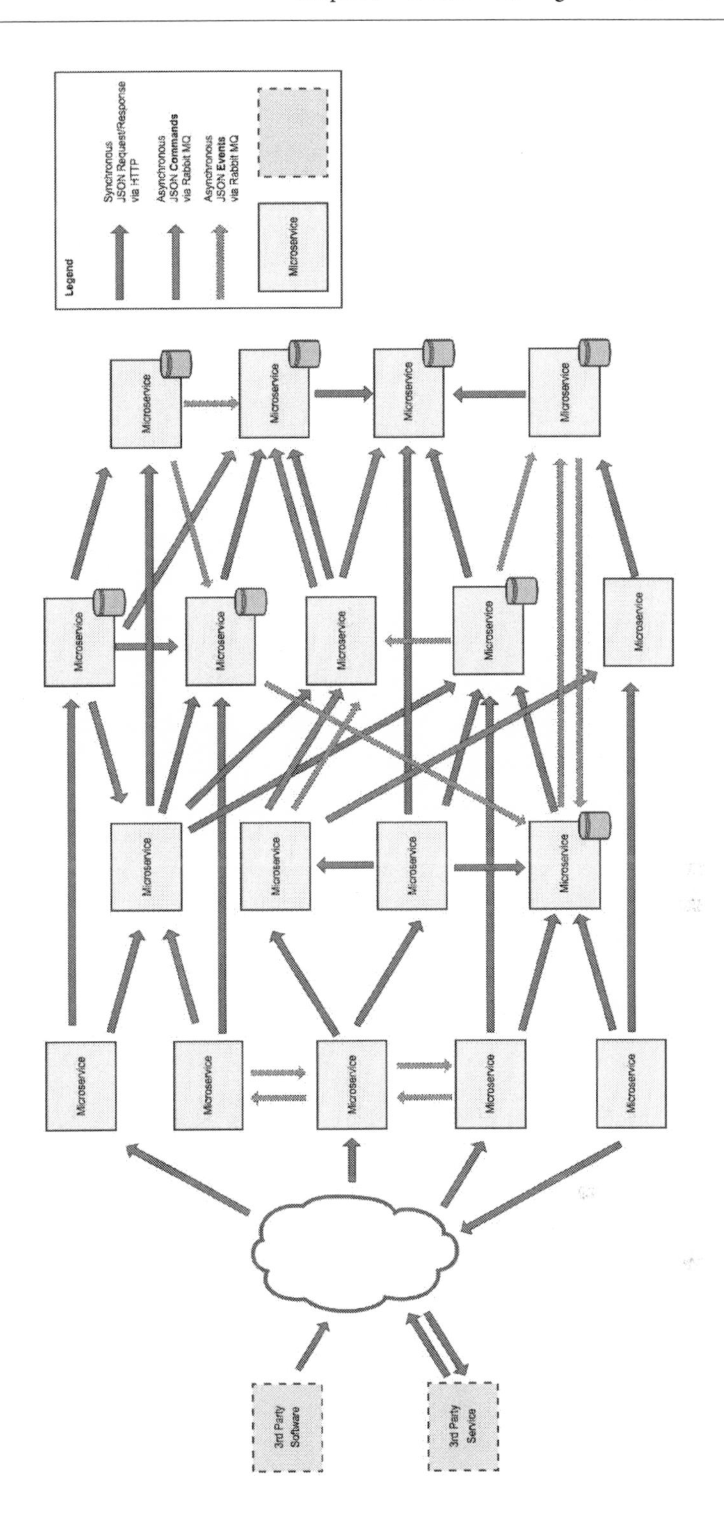
Legend
Synchronous
JSON Request/Response
via HTTP
Asynchronous
JSON **Commands**
via Rabbit MQ
Asynchronous
JSON **Events**
via Rabbit MQ
Microservice
3rd Party
Software
3rd Party
Service

> "a regulated institution would include IT security considerations throughout the software development life-cycle including requirements-gathering, design, programming, testing and implementation phases."

We couldn't provide the assessor with a document that described our software development process. That was because we had not documented them.

We also had no systematic way of addressing security in our application development, with every team having different skills and doing different things. The assessor found that some teams might think about security during the design and requirements-gathering phase, and others during the testing phase. Or sometimes security might only be considered during the implementation phase, such as when the engineering teams are looking at firewall rules. It really depended on the individuals within the teams; some were more security-aware than others. As a result, the assessor found that there wasn't a consistent approach between the engineering teams as to when security was considered.

So, when we received the assessment report, it came as no surprise that it reported a lack of a formalised and consistent mechanism for considering security throughout our software development life-cycle. It was sad reading. Along with the report was an action plan. It provided a road map of a set of recommended actions to take to make the necessary improvements in three time frames - immediate, short to medium, and long-term. One of the short to medium actions was to document and socialise the software development life-cycle and ensure that security is included in each phase and in the definition of done. A long-term strategic action item was to provide training to teams to ensure that they are able to conduct security threat assessments and understand the risks faced by the applications.

We saw value in implementing the recommendations and wanted to "shift security to the left". We researched industry best practices for conducting security threat assessments at every phase of the software development life-cycle and decided to go with implementing the threat modelling process.

The Story

So, we wanted to do threat modelling. Threat modelling is described as an iterative process, starting in the early design phases and continuing throughout the application life-cycle. This was exactly what we wanted. We wanted security considerations to be taken into account early in the feature design phase and to be reviewed constantly. If there were any threats, we wanted them documented so they would not be forgotten, and future developers would be aware of them. And if there were any threat mitigations or countermeasures in place, we also wanted them to be documented.

After researching the various approaches for doing threat modelling, we decided to base ours' on Microsoft's threat modelling approach.

The approach

We started by creating a document explaining how our threat modelling process would work and what the final output should be.

Our threat modelling process could be decomposed into three high-level steps.

The first step was to decompose the application in order to gain an understanding of the application and how it interacted with external entities. It involved creating use-cases to understand how the application was used, identifying entry points to see where a potential attacker could interact with the application and identifying assets and trust levels that represent the access rights the application would grant to external entities.

The second step was to determine and rank the threats. We used DREAD scoring for rating the threats. DREAD scoring defines a threat rating based on 5 categories (Damage, Reproducibility, Exploitability, Affected Users and Discoverability). A score of 1 to 10 is given to each of the 5 categories. The final rating value is obtained by adding all the numbers and dividing by 5.

The last step was to determine countermeasures and mitigation for the threats. This involved sorting the threats from the highest to the lowest risk and prioritizing the mitigation effort.

We created a template for the threat model document and below is an example for a fictitious application.

Sample Application Threat Model

Application Brief	Sample Application - Application Brief
Network Zone	DMZ
Database	some-database
Web GUI Front End	some-framework
Message Broker	some-message-broker
Comments	This application is different from standard apps. It will be doing x differently because of y.

User roles/Trust Levels

ID	Name	Description	How they will use system
1	User With Valid Login Credentials	A user with authentication details.	• view x. • update x. • create x. • delete y.
2	Third Part Vendor XYZ	Staff from third party provider XYZ	• view and edit user details. • create x.

Entry Points

Id	Name	URI and Description	Trust Levels	Security Implementation
1	Monitoring TLS port 8223	The artifact will be only be accessible via Rest Services over SSL for monitoring it will be restricted to Monitoring server.	(1) Ops Monitoring	Firewall rules are set to restrict access to endpoint on port xxxx
1.1	Overall Health	This endpoint is for checking overall health. https://<hostname>:xxxx/<app-name>/services/healthCheck/overall	(1) Ops Monitoring	
2	HTTPS Port	The artifact user endpoints will only be accessible via Rest Services over TLS.		SSL is enabled using xxx encryption
2.1	<Function>Controller	This endpoint is read-only and returns information for the Account of the Authenticated User https://<hostname>:xxxx/<app-name>/account /<account_number>/info	(2) User with Valid Login Credentials whose account number is the same as the account number of the User	Annotation has been added that verifies user is authenticated and is that the account number matches the account number in the path variable of the REST request.
2.2	<Function>Controller	This endpoint is read-only and supplies Transaction summary information for the Account of the Authenticated User https://<hostname>:xxxx/<app-name>/account/<account_number> /transactions/summary	(2) User with Valid Login Credentials whose account number is the same as the account number of the User	Annotation has been added that verifies user is authenticated and is that the account number matches the account number in the path variable of the REST request.

Assets

ID	Name	Description	Trust Levels	Security Implementation
1	Customers	Assets related to our customer		
1.1	User Login Details	The login credentials that a customer uses to log into the system.	(2) User with Valid Login Credentials	
1.2	Account Information	This application stores the account information including transactions and back account information.	(2) User With Valid Login Credentials (3) Ops (4) Sales (5) PROD Database Read User (6) PROD Database Read/Write User	Rest interfaces are locked by verified authentication. Sales access is restricted by ssl and can only be accessed via web interface.
1.3	Personal Data	The College Library website will store personal information relating to the students, faculty members, and librarians.	(2) User With Valid Login Credentials (3) Ops (4) Sales (5) PROD Database Read User (6) PROD Database Read/Write User	
2	logs	Assets related to log information stored	(3) Ops	
2.1	Request Log	Log stores information concerning each request.	(3) Ops	Logs are stored in a private network that can only be accessed by OPS

Threat Assessment

Threat	D	R	E	A	D	Total	Rating
Attacker obtains authentication credentials by monitoring the network.	7	7	5	8	9	7.2	High
SQL commands injected into application.	8	8	8	8	8	8	High

Threat Risks

Threat Description	Attacker obtains authentication credentials by monitoring the network
Threat Target	Web application user authentication process
Risk Rating	Low/Medium/High (derived from threat assessment above)
Mitigation Level	0 - Not mitigated 1 - Partially mitigated 2 - Fully mitigated
Mitigation Strategy	1 - Do Nothing 2 - Inform about the risk 3 - Mitigate the risk 4 - Accept the risk 5 - Transfer the risk 6 - Terminate the risk
Risk	Malicious user is able to login as another user and: • see financial details • perform financial transactions • view personal information • alter user information and account details • gather information that will allow user to attack other third party systems
Attack Techniques	Use of network monitoring software
Countermeasures	Use SSL to provide encrypted channel

How we introduced it

Trying to change a team's process from the outside is hard. To introduce the threat modelling approach, we first gave a presentation to the security champions. We have a security champion in every engineering development team. They are the most security enthusiastic person in the team and their role is to put on the security hat and ensure that teams are always considering security. The threat model approach was put together by a few security champions who were looking into what processes to include in our engineering practices in order to shift security to the left. Once we got feedback and support from the security champions, we presented our approach to the entire engineering team. We asked them to create a threat model for every microservice application that they owned. We suggested they start by creating a threat model for every new application and slowly work their way back and perform a threat model on all their other existing applications. We also asked the security champions to book threat modelling sessions with their teams and help facilitate every team's first session.

After the presentation, everyone was on-board with giving threat modelling a go. The teams who were about to create new applications reached out to us to help facilitate. The other teams made plans to threat model their newest application while its architecture was still fresh in their minds.

How we thought threat modelling would be done

During a threat modelling session, members of the team would be asked to draw the architecture diagram for the application under scrutiny on the whiteboard. From there everyone would work to identify the application's entry points, the user and roles that were required to access those entry points, and the external dependencies for the application. Next everyone would refer to the architecture diagram, along with the other data that had been listed out, to identify potential threats. Each threat would then be given a threat rating using DREAD scoring. Once the threats had been rated, the team would order the threats from highest to lowest. They would then work their way down the list, determining countermeasures and mitigations against those threats.

When the threat modelling session was over, the information that had been drawn on the whiteboard would be captured and someone from the team would volunteer to record all that information into the application's threat model document. We would then ask teams to review the threat model document every time a new feature was planned which involved that application as well as whenever code changes were made as part of a security review. That way the threat model document would always be up to date.

What actually happened

When we ran threat modelling sessions for new applications, we ran into something interesting. How do you decompose an application that has not yet been created? How do you define entry points and assets for something that does not exist? As we were developing using agile methodologies, requirements weren't necessarily set in stone at the point that engineering work started. The way we worked was lean and iterative and we only had user stories as our requirements. We knew what the application needed to do, and the design and architecture would evolve as we worked through the user stories. Not only were our requirements lean, so was our code. As mentioned, the practice of TDD means no production code is written without a failing test. And the test should not check functionality outside what is needed to complete the user story. We ended up with more placeholders than content as there were a lot of sections in the template that we were not able to complete during the initial threat modelling session. The teams said that they would fill them in as they developed the application.

When it came to determining threats, we found that as the applications were micro services, they all had quite similar functionality. The front-end applications would generally handle user input or receive requests from a third-party service. All applications would either send or receive messages to or from at least one other application and those in the back-end would usually have a data store. Rather than repeating the same code to implement all that functionality for each application, those common functionalities were extracted out into common libraries. With the common shared libraries, we had a standard way for doing things. Those libraries were all security hardened and had automated tests that verified that the security controls were in place. When an application used those libraries, it would also inherit the security controls and the tests which verified them. Those parts of the threat model were exactly the same for every microservice.

We also ran into some difficulties with threat modelling legacy applications. Being legacy applications, no one in the team fully understood all of the application's functionality. They were only familiar with the parts they had worked on. As such, we were not able to fully decompose the application. We drew the architecture diagrams and mapped out the entry points of what we knew and focused on identifying and rating threats for those areas. Because we used shared libraries, a lot of that was pretty straight forward. We cut and pasted the standard text we had used in other applications that also use the same library. The text would say something like "We will use SSL to prevent malicious users from snooping traffic. Use inter-system authentication. Access restricted to the x network". For the other threats that were unique to the application, we used DREAD scoring to determine the threat risk. We did not realise how hard that would be.

There were some threats that the team could not reach consensus on. Should we be rating the threat taking into consideration the application's location in the network? If so, did that mean that all internal applications, which are behind firewalls, should be rated low on reproducibility, exploitability and discoverability? What if an attacker manages to secure a foothold in the network? The security team decided that threats should be rated without taking into account any security controls external to the application.

This was a point of contention. Many engineering teams did not agree and there was always a lot of discussion around the rating for each of the DREAD categories. Team members were arguing about whether the discoverability of a threat should be a 6 or a 7. You wouldn't think it was a big deal, and I didn't think so either. However, those slight shifts in scores for each category return a lower overall score, which in turn gives a lower rating. A threat with a medium risk might have its risk downgraded to be a low. **DREAD scoring became really dreadful**. Teams had initially scheduled an hour for threat modelling. It usually took longer. The longest took three hours and that team did not even finish rating all the threats, let alone determine countermeasures or mitigations. They spent the majority of their time arguing on the threat risk rating. Everyone was frustrated at the end of the session. They could not reach consensus, they felt that they had spent too much time on it and could not see the value.

At the end of the sessions, the teams would agree that some pre-work needed to be done to document the rest of the application's functionality and entry points. Once that work was completed, they would schedule more sessions to complete the rest of the threat model. At least, that was the theory. Of course in practice, none of that eventuated. The truth is that nobody likes documentation. Threat modelling was tedious and boring. It was important, of course, so a task card would be written up and stuck on a team's board. And there it would stay, slowly bio-degrading, because nobody really wanted to work on it. Eventually some were selfless enough to pick it up, or they drew straws. This hapless developer would spend some time on it and give up in dismay after a few days as it was monotonous work. I don't think anyone can fault them for that. Some of the legacy applications have a lot of entry points; our internal web portal had over 370 entry points! I'm not even sure if that's the largest since the other legacy applications never had their entry points fully listed. I would not want to be the person having to go through that much code to do all that work. The teams also did not schedule any future threat modelling sessions for their legacy applications. They did not want to go through the frustration that they had already experienced as they felt that their time would be better spent doing other work.

Not long after we introduced it, the teams started having threat model fatigue and were avoiding it. Even if they had a good experience with a newer small application, once they tried to threat model a legacy application, it left them with a sour taste. Like that swig of cold coffee from last week you accidentally took because you forgot to clean the mug off your desk.

When we introduced threat modelling to the organisation, we also introduced a process of doing security reviews for every code change as a way for keeping the threat model current. That is another epic failure and the details for that are for another story. As a result, threat models were not updated when implementations changed. That made teams even more reluctant to spend any more time on them. They didn't want to work on updating a document that was already out of date. In the end, most threat model documents were left uncompleted.

Threat modelling was an epic failure in our DevSecOps journey. Teams scorned it and the term threat modelling became a dirty one among engineers. They could not see how threat modelling, in its current form, was going to make their applications more secure and felt that they were only doing it for compliance reasons.

Attempting to improve the process

Having identified some of the pain points, we made two changes to improve the threat modelling process.

The first change was to generate the list of entry points from the application's source code. This was done via code executed by the test framework that produced a table that could be copied into the threat model. What once took a couple of hours could now be generated within seconds.

(The code that generated an application's entry points)

The second change was to record the threat modelling outcomes in code. Keeping the threat model information next to the affected code would provide greater visibility for those looking at the code or modifying it. It would also make it much easier to keep the information current.

We created annotations for defining assets, security implementations and risk assessments. The screenshot below shows an example of how the threat modelling **Asset** and **RiskAssessment** annotations were used.

```
//Imports
...
@WebAppConfiguration
@ContextConfiguration(classes = {MySampleApplicationContext.class, MySampleApplicationWebContext.class,
MessagingTestConfig.class})
public class ThreatModelGenerationTest extends AbstractJUnit4SpringContextTests {
@Autowired
private RequestMappingHandlerMapping handlerMapping;
@Test
//@Ignore
public void printEndPoints() throws Exception {
ThreatModelUtil threatModelUtil = new ThreatModelUtil();
threatModelUtil.printTrustLevels(handlerMapping);
threatModelUtil.printEndpointsWithThreatModelConfluenceMarkup(handlerMapping, 1);
}
}
```

```
@RestController
@Validated
@Asset(name = "xxx Details", description = "Details of the asset goes here", value = Asset.AssetType.USER_DATA)
public class SomeController {

    @PreAuthorize("isLoggedIn(#someValue)")
    @RequestMapping(value = "/someEndPoint/{someValue}", method = RequestMethod.SOME_REQUEST_METHOD)
    @RiskAssessment(description = "Attacker some do something really malicious",
            target = "some process",
            Level = "4.4",
            mitigationLevel = "2",
            strategy = {"3",
                    "We will use xxx to prevent malicious users from snooping traffic",
                    "We will use xxx to prevent sniffing, and replay",
                    "Database is kept in the xxx network and does not provide any access to xxx or external traffic"},
            risk = "Malicious user is able to login as another user and:\n"
                    + "* see xxx details\n"
                    + "* perform xxx transactions\n"
                    + "* view xxx information\n"
                    + "* alter xxx information and xxx details\n"
                    + "* gather information that will allow user to attack other third party systems",
            attackMethod = "Use of network monitoring software",
            countermeasures = {"Use xxx to provide encrypted channel",
                    "We will use xxx to prevent sniffing, and replay"}
    )
    public SomeVaribleType getSomething(@PathVariable @NotNull SampleVariableType someValue) throws Throwable {
        return someService.getSomething(someValue);
    }
```

With both these changes, most of the threat model's output could be automatically generated. We were excited with the enhancements and could not wait to share them with the developers.

The unexpected response

When we presented our automated threat model enhancements to the engineering team, we were expecting them to have a change of heart towards threat modelling.

Instead, they reacted to our improvements with disdain! "You've tainted our code!", they said. "Look at the size of those annotations!". The threat modelling information usually spans over a couple of lines and was blowing out the size of our classes. It also took up too much screen real estate which would slow down development. They felt that it would become a disincentive to have any useful information in a threat model or even to keep it up to date.

We were gutted. We had invested time into those improvements hoping it would encourage threat modelling. Instead, it had the opposite effect. Lessons learned: we should have consulted with members from the development team before investing so much effort into something we thought they would like.

A retrospective

By the time we tried to introduce those "enhancements", it had been over a year since we started threat modelling. There were still a number of applications without a completed threat model and probably a lot more which were outdated. It didn't feel like threat modelling was improving our security posture and it definitely wasn't shifting security to the left. In fact, it felt like threat modelling was making teams want to avoid security.

We held a retrospective with the engineering teams and also engaged an external consultant to help us determine if there was a better approach to help shift security to the left. The result from the retrospective, along with the findings of the external consultant, revealed that teams felt that there was a lot of repetition in the threat modelling document. All our applications use a number of shared libraries which are already security hardened. Therefore, they are secure by default. Teams did not want to be spending time documenting security controls which were standard across all applications, they would rather it all be automated. They would rather spend their time thinking and discussing mitigations and countermeasures for threats not already covered by those libraries.

A different approach

We decided to change our entire approach to threat modelling. We changed it to be a suite of unit tests. These tests don't just document a particular security control, they also validate that the control exists. That aligns very well with Test Driven Development. We already use security stories when planning features. The security stories help identify any security work that also needs to be done when building a feature. Developers can then implement a security test as part of that work to test that the controls are working correctly. And because we have a continuous build pipeline, those tests are run every time an application is being built. So, we are continuously validating that our security controls are still in place. At the end of the build pipeline, if the application has passed all the tests, a new threat model is automatically created and published.

We spent some time creating a test framework for this and then invited teams and security champions to work with us to develop some of the common security tests. Below is an example of a security test that checks that the security configurations for a web framework are correctly set.

```
@Test
public void shouldVerifyThatCrossSiteScriptingAttacksArePreventedUsingHdiv() throws Exception {
    SpringRestEndpointDefinition springRestEndpoints =
            new SpringRestEndpointDefinition(applicationContext);

    securityTestFramework.applySecurityStory(
            (SpringRequestMappedEndpoint applicationBoundary,
             VulnerabilitiesDescriber vulnerabilitiesDescriber) -> {
                RequestMappingInfo requestMappingInfo = applicationBoundary.getRequestMappingInfo();
                Set<String> pathPatterns = requestMappingInfo.getPatternsCondition().getPatterns();
                Set<RequestMethod> methods = requestMappingInfo.getMethodsCondition().getMethods();

                if (hdiv.containsExcludedRequest(pathPatterns, methods)) {
                    vulnerabilitiesDescriber.raiseVulnerablity( ... );
                }
            }
    ).toEndpoints(springRestEndpoints);

    securityTestFramework.verify();
}
```

If the security test failed, it would throw an error and break the build. The output of the error was security-specific and provided a suggested fix.

Below is a sample error output:

Vulnerability:

Key: missing-hdiv-xss-protection:com.company.some-app.controller.SomeController#some-request#5bbg13e-a59a11a5046755d6b8021be31
Location: /someEndPoint/someAction POST
Description: Hdiv configuration permits Cross Site Scripting vulnerability for /someEndPoint/someAction POST
Suggestion: Remove Hdiv exclusion from configuration for this endpoint

If teams want to accept the risk, they can perform a risk assessment and add it to the application code using Domain Specific Language (DSL). The risk assessment would prevent the security test from failing and breaking the build. The main difference with this new approach is that a risk assessment is only needed if the threat is not going to be fully mitigated. This is not often the case, so teams are happy with that amount of documentation.

Below is a sample risk assessment.

```
ThreatModelDsl.riskAssessments {
  riskAssessment {
    description "Some threat description"
    suppresses "missing-hdiv-xss-protection:com.
      company.some-app.controller.SomeController
      #some-request
      #5bbg13ea59a11a5046755d6b8021be31"
    risk {
      impact MODERATE
      likelihood RARE
    }
    mitigationLevel RISK_ACCEPTED
    mitigationStrategy "Some mitigation strategy"
  }
}
```

This new approach was very well received by the teams. They were happy to work together with us to create the common security tests. One team discovered a few security test failures when they started using the tests and were able to quickly rectify the problems.

We also discovered an added advantage to this new approach. We were able to create new security tests based on findings from our security penetration testing. This meant that all the applications were able to benefit from those findings. We were able to identify potential security issues earlier and address them before they went to production. As a result, the number of duplicated security findings decreased. We were truly shifting security to the left.

Lessons learned

We took away a couple of learnings from this. We learned that in order for threat modelling to be successful, we needed to do three things. We needed to demonstrate its value to get buy-in from developers to spend time on it. We needed to get early feedback from them to ensure that it will work for them, and we also needed to automate as much as possible to reduce the extra time they will spend on it.

When we first introduced threat modelling, we shared with everyone what we wanted to achieve, which is to shift security to the left, and how threat modelling would get us there. Everyone was on-board with the idea and was keen to try. However, they felt that the initial approach was impractical. The DREAD scoring was dreadful and the repetition of the same controls for each application unnecessary. In short, they did not see the value in that approach. Nor did they see value in creating and maintaining documents that captured default security controls. Rather, they found value in having discussions about security concerns that hadn't already been covered by the defaults. The new threat modelling approach did just that. It removed all the repetition and allowed them to focus on the security concerns that truly matter. The teams were able to get behind the new approach because it demonstrated its value.

Getting early feedback is also key. When we started, we did not involve the engineering teams. We had a small number of security champions working in isolation to design the threat modelling approach. Although the security champions were also engineers, they represented a small subset of the engineering teams. With the benefit of hindsight, there were many issues with the initial approach that are quite obvious now. If we had involved more teams from the beginning, we might have discovered those issues earlier. I believe one of the reasons the new threat modelling test framework approach worked is because we obtained early feedback. We invited teams to join us and we were able to build something that worked well for them. The new approach was very well received as it aligned with the agile methodologies that we adhere to. If we had not gotten early feedback, we might not had been able to make that connection. And the new threat modelling approach might not have been so successful. Therefore, early feedback is key!

The last lesson we learned is to automate as much as possible. I feel that this is especially true when it comes to security. The more you can automate and remove the friction, the more likely it is to succeed. In fact, automation for simplification, repeatability and speed is a key ingredient for "shifting security to the left". Those we wanted to be involved in shifting security to the "left" already have lots to do and the only way to get them to look after more is to automate more. One thing about the new approach which I'm most proud of, is that the threat model documents are generated and published after every successful run in the build pipeline. The threat model is always current, and it documents not what we think or want the application to do, but what it actually does. The unit test also automatically checks that the security controls are in place and provides fast and early feedback to developers. Doing so allows them to develop quickly and securely. Automating security, especially in the build pipeline, has helped make it everyone's responsibility and is an important part of our DevSecOps journey.

Even though our initial threat modelling approach was an epic failure, the lessons we learned were invaluable. We've learned that we need to demonstrate value, get early feedback and automate as much as possible.

We've since applied those learnings in every security initiative we're introduced.

The trouble with documentation is...

... that it always ends up out of sync with reality.

Many thanks to Simon Gerber who gave me permission to share an internal wiki page he wrote about his company's attempt to label the cutlery holders.

Somebody updated the documentation ...

Didn't help.

About Edwin Kwan

Edwin is the Application and Software Security Team Lead for a bank. His approach toward application and software security is to raise security awareness, provide light touch controls to the software development life cycle to increase visibility of security issues and work closely with engineering teams to quickly develop secure applications.

Edwin started out as a software engineer and transitioned into the application security role to lead a range of security initiatives when the company was working towards obtaining an unrestricted banking licence.

As a Software Engineer, he has over a decade of experience developing large scale; real-time; high performance; high reliability software applications for major telecommunication vendors. He is also experienced in working with stakeholders from small to large organisations to design and develop innovation solutions to help manage and grow their business.

Chapter 6

Red Team the Culture

by Fabian Lim

Chapter 6
Red Team the Culture by Fabian Lim

Introduction

At first, few in the organization knew what DevSecOps was. But after seeing how DevSecOps can benefit my organization, the tech leaders and directors immediately wanted all engineers to embrace it. One principle in DevSecOps embraces the mechanism to conduct tests (and breaking your own products) with the intention to be better, faster, safer. This method is sometimes known as 'Red Teaming'.

'Red Teaming' is not just a glorified penetration test (a.k.a. pentest). It involves much more work BEFORE and AFTER the pentest. In a typical pentest, the rules of engagement are defined prior to the engagement and are usually limited in scope or boundary, to specific applications, machines or a certain technology. A 'red teaming' exercise differs from typical pentesting as there are fewer boundaries or rules. It often simulates a real cyber attack and could extend the scope to physical sites, humans or any other method or techniques this red team could engage with. However, the most crucial part of 'red teaming' engagement is the remediation. The remediation offered by the Red Team must be in context and relevant enough for the developers to understand and fix the issues easily.

From the perspective of a DevSecOps engineer working in the trenches, this is a story about how 'red teaming' brought about a change that is more than just technical but also the start of a long-awaited cultural transformation as we engaged our own Red Team to perform a 'red teaming' exercise.

This journey will show how a collaborative effort between two teams can turn sour quickly as it was encouraging and delightful initially but became a relationship disaster at the end. We experienced various push backs, hurt feelings and even the loss of trust. We wished that the benefits from 'red teaming' were as good as promised. Even with these issues, there were positive outcomes and lessons learned, although the price tag might have been too high.

The Story

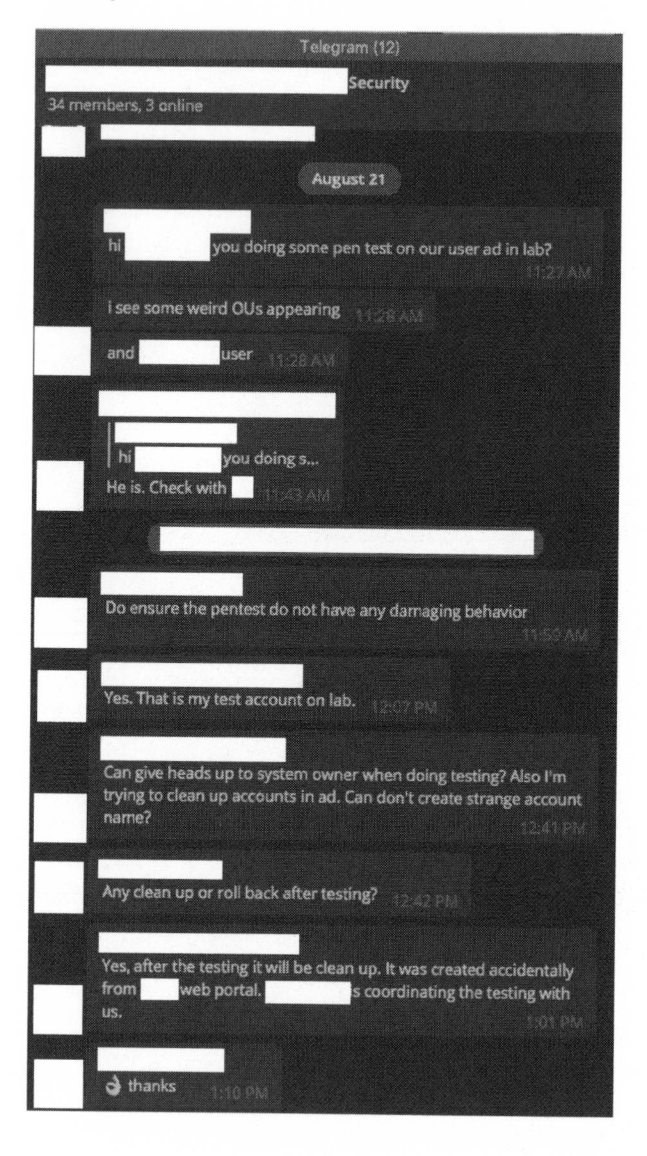

"Do ensure the pentest do not have any damaging behaviour."

"Can you please give a heads up to system owner when doing testing?"

"Don't create strange account name."

"Please clean up after yourself."

These are some reactions by my colleagues following a recent in-house penetration testing (a.k.a. pentest) exercise on our systems by us, the product security team. I thought everyone in the product team was cool and collaborative but somehow these bold efforts are still frowned upon despite being open about our objectives in the JIRA ticket. The mention of "pentest" will have gotten you the notorious reputation.

I am part of the above mentioned one-pizza-size product security team within the larger product team. I am also a developer in the product team so my world is a union of security and DevOps - DevSecOps. A year ago, due to inadequate manpower and resources within the product team, we leveraged resources in another department, a separate Red Team, to perform a 'red teaming' exercise on our product team. The Red Team was formed recently and consisted of cybersecurity experts with various specialized skill sets. They were also looking for their first engagement to refine their 'red teaming' processes. So, after two years of peaceful product design and development, our product was months away from launch and this exercise can also fulfill the compliance check box. It was thought to be a win-win situation.

Developers were used to annual pentests, but this 'red teaming' exercise was something novel to the team. Some of us thought that it is just another pentest but it is not. For the most part, pentest deals only with analyzing vulnerable code, machines, and servers. However, 'Red teaming' exercise expands the scope and assess human-related vulnerabilities like our processes, best practices, behaviours and enforcements, etc.

I knew what I signed up for when I agreed to it - a subject in the exercise as I am also a developer. This exercise was announced to the rest of the team a few days before it was conducted but no details were given. Everyone was supposed to operate business-as-usual but also be ready for any 'red teaming' activities. And so it began...

The 'red teaming' exercise included but was not limited to, social engineering methods like phishing, gaining physical access to machines, and also installing malware on laptops.

The red teamers' successful social engineering attempt got them physical access into the building. They had disguised themselves as some kind of external auditors doing audit checks on our environment. The red teamers also managed to convince the developers to open an 'innocent-looking' document file from an unknown thumb drive which installed malware in the background that established a program that communicated to a command and control server - which the red teamers hosted - to maintain post-exercise control over those compromised laptops. From there, they managed to get constant and direct access to some of our developers' laptops, which allowed them to obtain source code and other sensitive information, like IP addresses and credentials, in the lab environment. These infected machines were then used as pivot points to other machines in the lab and other unsecured machines in the same network.

Everyone, including myself, was unaware of all these events as they were happening. I admit I was a little jealous the red teamers managed to pull off all those attacks and it was like a scene from 'Mr Robot' that turned into reality.

Our report card was out and our grades were not surprising at all. We knew our shortcomings and the report was just another avenue that stamped it on paper. In addition, the exercise discovered more flaws than we would like and we failed the exercise. In this meritocratic society, failure brings bad reputation and often leads to scrutiny.

The main reason why the 'red teaming' exercise was successful is because we were too ready to trust authority but did not seem to take the extra step to verify. We took it for granted that anyone who walked through the 'secured' glass doors was authorized, let our guards down and submit to authority without challenging their authenticity. Unaware, our developers submitted their laptops to the 'auditors' for checks.

On top of that, there were other behaviours that were bad practice that were frowned upon. Some laptops were left unattended without screen lock for an extended time, a common user and password were used across almost all applications and machines - even privileged accounts, etc.

The red teamers had slipped under our radar and pillaged our resources. We did not detect or report any suspicious activities.

The Outcome

From the exercise, it is clear that the Red Team's plan was well organized planned and beautifully executed. They did their research well - they knew our environment and also knew their targets.

However, the relationship between the developers and the Red Team turned sour in the aftermath.

For most of the team, we were shocked not by the results, but by the unorthodox methods used during the exercise. The exercise came as a surprise for many, stirring mixed feelings. The product team responded with a multitude of push backs - some took it personally and were upset that the exercise had 'invaded their privacy' (even though the laptops were issued by the organization as a development tool, never a personal device); some said that it was breach of trust between them and the Red Team; some will 'never trust the Red Team again'; others thought 'the Red Team should never have done this to us'. But few thought that this exercise was worth the effort - we had extra help to do security checks. It is better late than never; better us than the real bad guys.

Part of the reason why some felt this way was because we had personal information, like documents and credentials, and other social media profiles on the work laptop. People also grow feelings toward their laptops. Some thought that the methods used during the exercise were not appropriate because now it made us feel paranoid and insecure about everything we do. We felt the tension in the air as everyone is uptight about the event even after the managers tried to explain the exercise's objectives and address any concerns during retrospective team meetings.

As part of post-mortem investigation, the product security team was tasked to collect a snapshot of the image of the compromised laptops for forensics purposes and then restoring them to a known good version. However, some developers were not keen to do so. Handling matters in their own ways, they formatted their compromised laptops on their own accord in frustration to get rid of the malware and restore the laptops to their desired state. In doing so, they denied the product security team from collecting those snapshots required for post-mortem investigation. In their defense, there was no formal post-mortem procedure and enforcement at that time and developers were not educated or never knew they were subjected to such procedures.

Within the few weeks after the exercise, we worked with the Red Team to tabulate the findings and analyzed the security risks. This final, joint report was very much anticipated by the higher management as it was the first of its kind and also not a typical independent pentest report. But this also raised problems.

The Red Team wanted the findings reported as-is, immaculate and unmodified, complete with Common Vulnerability Scoring System (CVSS) ratings - no different from a traditional independent pentest report. The product team was concerned with some of the wordings which would make us look bad so we wanted to add justifications to the findings cushion the effect with a slightly less serious tone. There was a struggle between the two parties because both sides had their own agenda to fulfil. To me, it feels like another paper-pushing war as no one was willing to compromise.

In the report, the product team wanted to highlight that this exercise was conducted entirely in the 'lab environment'. A struggle that I have was with the so-called 'lab environment'. If source code were to be developed and tested in this 'lab environment' only to be copied into the production environment, shouldn't we also secure this 'lab environment'? In a similar way, when a nation develops any secret program, the development 'lab environment' of this secret program is also secured and classified. They take caution that they do not expose any evidence at all. So why should we?

But because of this magical, untouchable 'lab environment' label, some tried to cushion or downplay the effects of the findings by rationalizing that this exercise was performed in the 'lab environment', which does not reflect the real production environment. The findings are only a 'cry wolf' phenomenon.

The paper-pushing went on for more than a few weeks and there was no agreement. The pressure from the higher management was even more intense than before, and in order to meet their demands, the Red Team sent their report as-is.

This soured the relationship even further. Now, the distrust between the product team and the Red Team reached a higher level. We were told by our product team leads to 'disengage the Red Team' until further notice.

The product team quickly responded and followed up with our version of the report - with mitigations, risk assessments, security plans, and justifications. And of course, this went back and forth with the higher management a few rounds before someone decided that this report was somewhat acceptable.

The product team knew that the findings had to be fixed before 'something major hits the fence'. Our approach towards security was to strike a balance between development practicality and calculated risk - rather than compliance to existing policies. We want to do what is the best for the product and for us. With this intention and approach, and using the report as a reference we started to convert the plan into tasks as we know how - creating JIRA tickets in our product backlog and started to prioritize them so that it will eventually get done in our sprints.

In order to complete the long laundry list of security tasks, we set aside a fraction of our man-hours every sprint towards fixing security-related bugs. Trying not to repeat the mistakes, we first fixed our test codes. We approached new product features in a different way that had reviews which involved subject matter experts, like the folks with security expertise, to review and aid in designing secure workflows and models from birth rather than at the end of the implementation. We still struggle with managing the priorities of security bugs versus new product features. But we definitely placed more emphasis on having secure design and code.

We also managed to identify and fix some gaps in our processes so that we can mitigate some of the security risks we previously overlooked. For example, our development laptops are now monitored under a mobile device management (MDM) program.

Apart from technical debt, the cultural debt was the hardest to fix. Since the 'red teaming' exercise, everyone started to exhibit some vigilant behaviours. I am unsure if it is a knee-jerk reaction that subsides after a while or if it will continue. So I started a mini 'lock your laptop' campaign within the team to (constantly) remind ourselves to exhibit good habits within the work environment. Whenever I see an unattended unlocked laptop, I would use the victim's laptop to log in our team's chat group and make a random attention-grabbing announcement. I try to do this without shaming or making them look too silly.

Apart from this seemingly silly constant reminder to lock laptops, we had minimal help towards improving the culture and behavior. Perhaps we did not know how. It felt like we were only slightly better than before. The cultural debt still requires more work to get to where we want to be to operate effectively and efficiently as an organization.

On the bright side, we had requested and received more resources - manpower and funding - from the higher management to address the above technical debt. We took slightly more than a year to fix the issues. Now we are in a better shape than before - our code stinks less and there seems to be a slight shift in culture towards being more open to the general idea of "self-testing" and "being a target for testing". I look forward to the next 'red teaming' exercise that will hopefully make another slight shift towards the DevSecOps transformation.

The Lessons Learned

Fix the cultural debt

The most important lesson for me is that a cultural transformation requires much more than a technical or a process change. It requires everyone to be part of the transformation and there is no one-size-fits-all solution. In this case, the organization had introduced a new process ('red teaming' exercise) as part of adopting the DevSecOps transformation when more consideration could have been given to the human and cultural aspects of the transformation.

Cultural debt hurts everyone. The developers, who are at the end of the chain, are frustrated as a result of the hidden agendas and political warfare not within their control. As a result, our productivity tends to drop.

Loosely speaking, culture essentially defines who we are and how to think and behave. By encouraging positive behaviours and habits in our daily work life, it will help shape a better culture. We should always encourage each other to keep up with the good practices, and sound out and improve bad ones. The mission and vision have to align and resonate with everyone. It is key that the business must align with the development and security teams to move towards the organization's mission.

The fact that 'red teaming' had caused a slight shift in behaviour might mean that we have to do it repeatedly until the point that it becomes a norm. The Red Team might just be the catalyst that kick-started our transformation. If I had control over the pace of 'red teaming' exercise, I would make it happen often.

Align and prioritize with a security point of view

We managed to fix security issues as it comes because the team aligns and prioritizes security in sprints to make sure security moves at the speed of the product. This could not have been possible without the team and the product owner placing emphasis on security fixes and features. We embraced Agile principles and were using the SCRUM framework. It was clear that these require attention to avoid security debt in the future. It is better late than never; and better our whitehat red team got to us first rather than a real black hat attacker.

Build a trusted relationship with the Red Team

We had no clue who the Red Team was and how it was going to be conducted clearly. We were notified with an email saying: "There will be a 'red teaming' exercise for this two weeks. Be prepared." Without any advice, we were clueless about what we should and should not do. We were clueless about the intentions of this exercise.

A successful 'red teaming' exercise requires having a trusted relationship between the Red Team and everyone. The building of relationships take time and effort and it involves everyone to make it work. We should feel that we are able to trust the Red Team to uncover the deepest, darkest secrets in the code that is out of reach for other independent testers.

One way to build this trust is to establish clear communication about what the intentions and objectives are of a Red Team and the things they are allowed (e.g. copy the source code, etc.) and not allowed (e.g. copy personal pictures, etc.) to do. This should alleviate some fear away from the targets and have a peace of mind that our privacy will be kept confidential.

A reminder on some of the best practices and behaviours should be communicated to ensure that everyone is on the right track before a 'red teaming' exercise. While the team leader of this exercise cannot manage all expectations and reactions (because we are after all humans), we should minimise clashes, emotions, and friction.

After an exercise, there should be no blaming in this relationship as the main purpose is to help the organization to move towards a better security posture. We should keep communication open, get feedback constructively and give due credit to all parties.

Action and solution speak louder than words

A common mistake people tend to make is to tell others how bad their code is but fail to offer any constructive feedback. We should bring solutions to the team, not just the problems. As Ian Allison mentioned in his blog post[1], the most important part of a red team engagement is remediation.

The Red Team did not follow up with contextual or useful remediation. It felt like they were an arm's length when it comes to remediation. It felt like they did not care to provide any relevant solution for the developers and expected the developers to find solutions for ourselves. They dropped a bomb on us and left. More often than not, developers do not understand the security jargons that security folks use. It is no wonder that developers had such negative reactions to the exercise.

It is important that people with both security and development knowledge work closely with the developers by communicating constantly, openly, clearly and directly. This communication can be facilitated through a common ticketing or messaging platform that both the developers and security folks use instead of sending emails and reports of 'security speak' nature.

Security folks may have to deep dive into the code and understand developers as much as developers need to understand security terms and risks. By working together regularly, folks earn other's respect and help each other. This way, we fix not only the bad code, but also promote overall good coding behaviours.

How should success look like

'Red teaming' is effective in an organization that truly has an open culture and acceptable to failures and feedback. In order to engage in successful "red teaming" exercises, the organization must be open to embrace all that comes with it. Red teams should also be risk-taking and push or even break all kinds of (cultural, technological) boundaries. Merely having a red team without any cultural transformation

1 http://www.devsecops.org/blog/2015/12/10/red-team-pwning-the-hearts-and-minds-one-ticket-at-a-time

will only suppress its ability into 'yet another pentest team' and not bringing it to its fullest potential.

When 'red teaming' is conducted fruitfully, the feedback loop is constant, open and well intended. Security folks would suggest and contribute resources (links, code snippets and pull requests) in the code repository or ticketing platform that facilitate the remediations. Developers welcome and accept security issues as a normal bug in the software. Managers willingly prioritize security bugs accordingly to risks and understand the risks behind each security ticket. Everyone should be moving in the same direction according to the vision. And then, we can almost be sure to say that that is when DevSecOps has successfully shifted left and transformed the organization.

In order to reach this stage, it takes a lot of effort from engineers and non-engineers; security and non-security trained folks; managers and non-managers. There will definitely be friction at first but it is our job as DevSecOps servants to be the bridge between security and non-security trained, to smoothen this transformation process.

Thank You

Special thanks to Pieter Danhieux, Doug McDorman, Stefan Streichsbier, Edwin Kwan, Mark Miller for all the editing and publishing efforts on this chapter and book. You guys made this world a better place for me.

About Fabian Lim

Fabian is a DevSecOps Servant, trying to make the world a better place. He has a deep passion for both technology and security - and hopes to help make the two combine seamlessly in the digital world. Fabian also likes to contribute to blogs, conduct workshops and conferences such as DevOps Days, All Day DevOps, DevSecOps Days with the aim to reach out to local communities with an interest in cyber security. His employment experiences include working for the cloud security team at Intuit Inc in San Diego, USA, where he first began his DevSecOps journey, and also for the Nectar team in Government Technology Agency of Singapore (GovTech). More on https://about.me/fabian.lim.

Chapter 7

Unicorn Rodeo

by Stefan Streichsbier

Chapter 7
Unicorn Rodeo

It was the largest and most cutting-edge project I've ever been on. Security was my responsibility. The application was very high-profile; major security problems were not an option.

On my first day, there was one squad and one microservice. Within a few months, there were dozens of microservices and a team of over 120 people.

You might wonder why I chose the title "Unicorn Rodeo" for this chapter. To me, this is what DevSecOps is truly about. Security seems at odds with DevOps and tech start-ups. How can one beat the competition with all these security roadblocks in the way? Modern security engineers have to be up to the challenge of enabling the business to move fast and be safe. Take a deep breath and mount the beast. Once you are on it, it will kick, buck, jump and do everything it can to throw you off. Two things can happen next: You fall off, or you stay on. If you stay on, eventually the unicorn will tolerate you.

This is a story of falling off a unicorn. But not for lack of trying to stay mounted. My goal in sharing this story is to help fellow DevSecOps engineers avoid the same mistakes I made. And who knows, maybe you'll manage to stay on it, in it and with it!

Yeehaw!

The Story

"This is going to be interesting", I'm thinking to myself as I make my way to the elevator lobby. I had been asked to help secure the development of a very high-profile mobile banking application. This is my first day.

Excitement is buzzing in the air as I got out of the elevator and walked towards the project team area. The project team occupied a quarter of an entire floor. Everything looked brand new. In fact, maintenance work was still ongoing which added to the vibe. This is how I imagined a well-funded startup, somewhere in Silicon Valley, would look like. Tables are arranged by squads, with cool sounding names. Good looking user journeys are all over the place, white-boards everywhere, and more post-it notes than sand on a beach.

It feels very unicorny.

Funny thing is, I'm nowhere near America, and most certainly not in a tech startup. This is a bank, in South East Asia, but it looks nothing like the other banks here. Not by a long shot.

I have heard that the people on this project are experts from all over the world and have an interesting challenge to solve. Re-write an entire existing application from scratch, and in a mere nine months.

One of the project managers spots me and waves me over. He introduces me to the lead architect and returns to his desk. After exchanging pleasantries, we dive right into the architecture. Not only did the place look like a Silicon Valley startup, the technology stack did too. What I've been mostly seeing in banks are Java monoliths. This was far from that.

Let's get into the juicy details. The main programming language was Javascript. Node.js using the Hapi framework for the back-end. React-native for the frontend, which included iOS, Android, and the web application. The Node.js backend consisted of stateless microservices that heavily rely on JSON Web Tokens. At a later stage, they would even add GraphQL. It doesn't get more bleeding edge than this.

At this point, I'm ecstatic. I love learning new things and this team had a great mix of experts and some new technologies that I'll get to dive into deeply.

However, as the senior person in charge of securing the application, I realized that it was not going to be easy. Challenge accepted! I was energized and walked over to the backend squad that was developing the first microservice.

The tech lead is great, very knowledgeable, and we paired up to go through the code. I immediately spot a few security integration points; such as Hapi comes with a solid input validation library. I also see a few code patterns that make my hacker sense tingle. I make a mental note and am keen to dig deeper. I thank the tech lead for the introduction and ask for help to get access to the code and set up my local development environment.

I'm positively surprised when I get the requested access in a few minutes. This team really gets stuff done quickly, I like it. I clone the repository from Gitlab and follow the README.md instructions. The local development environment comes up in a Docker container and just like that I'm all set.

Now that I'm up and running, I go to town on that microservice. After a few minutes, I confirm my suspicions from the pairing session with the tech lead. Most of the endpoints are indeed vulnerable to what is called an insecure direct object reference. It's a simple issue, that requires changing an identifier. This vulnerability is very common and fairly trivial to identify. Trivial to identify for humans. Security scanners typically cannot identify it.

I head back to the tech lead and ask for a moment of her time to walk through the first issues. She immediately understands and we discuss a fix for the issue. After all is clear, I ask her how I should file this in their JIRA. She gives me the details, I request access to JIRA. The first tickets are created and assigned shortly after.

In the afternoon, I have a session with the automation expert, who ran me through their pipelines. I was impressed with the amount of good work they had done on this. Their continuous integration practice is mature. Changes on certain branches would be directly deployed in the non-production Openshift environment. Deploying changes to production is also fully automated but requires a manual approval as a safety measure.

I can already picture integrating security tools strategically into the pipelines. It is all coming together nicely in my head and I feel fairly confident.

"Not a bad start.", I think to myself, "Not a bad start at all.".

The first week continued in a similar fashion. Things were under control, I was able to keep up with the pace of one squad easily. On Thursday I got asked to join a spike on the topic of Identity Management (IDM). The authentication logic of the application to date was mocked out and wasn't functional yet. Because the pace of development was fine, I had no issues joining the IDM spike. This spike would eventually turn into a major feature that required a lot of security design and engineering efforts. It ended up taking up a decent amount of my time. Try implementing a mobile banking application using stateless microservices. That's a very difficult engineering challenge indeed. But I digress, this is not what this story is about.

Most of my time was dedicated to designing the flows and use cases for the IDM. While being occupied, something interesting happened in the following weeks. Thinking back, the most accurate analogy would be that of the 'boiling frog'. I was in tepid water which slowly started to boil. Over the span of a couple of weeks the team size grew dramatically. Another squad was added, then two more, then four more. Instead of the one microservice, suddenly there were dozens under active development.

What started as a walk in the park, quickly became overwhelming. "How can I possibly keep up with them?" It was obvious that I needed more support. There was no way to catch up with the continuous progress made by the large and growing development team. I raised a request to create my own security squad and add three team members to it. It was approved quickly and allowed me to distribute tasks to my squad members.

The development team had a fast-paced development approach. That meant changes were made to all parts of the application on a daily basis. Furthermore, finishing the review of one microservice didn't mean you were done with it. This became very clear one Thursday afternoon.

Remember the very first microservice where I found issues on my first day? Yeah, that one. Guess what happened. The issue in the authorization logic, that was previously fixed, was re-introduced.

After creating a few microservices the team realized the difficulty of maintaining changes across them all. Key functionality was abstracted into Node.js modules. This triggered a refactor of all existing microservices. As part of the refactoring exercise, the fix became "unfixed".

A significant amount of time had been spent on improving the security of the application. However, I wasn't confident that the reviewed code was still in that secure state. Source code was continuously created at an ever-increasing rate. With an ever increasing team, working on an ever-increasing number of microservices. We couldn't rely on the same approach that we had for the last couple of months. We had to find a way to work smarter, not harder and faster.

"How can we guarantee that the entire application is in a good, secure state? And even more so, how can we verify that state on a continuous basis. Ideally across application environments."

Before coming up with solutions, we had to get a better understanding of what we wanted to achieve. These kinds of challenges are great and really get me going. I sat down in a quiet spot with a whiteboard next to me and started laying it all out.

First of all, we want to catch critical security issues as quickly as possible. For the Node.js tech stack, this meant getting security tools into the pipelines. As always, quick results are preferred. That meant open source security scanners were the weapon of choice.

The Node Security Project had made wonderful contributions to the open source community. Because of them, there were some decent tools out there that got us started quickly. As with all open source tools, they require some tuning to not be too noisy. Fortunately, this didn't take long. We were quickly able to establish a minimum security baseline. At that point, we were covered on common security issues in Node.js code. Additionally, we were alerted if any third party libraries contained known security vulnerabilities.

As it turned out though, both security tools were not providing much value. At least not for this specific application. The microservices didn't use any dangerous Node.js functionality, which is typically the cause of vulnerabilities. Also, dependencies were properly specified in the package.json. This led to dependencies being updated often, because of the very frequent builds. Any known security issues would be fixed almost instantly after a fix was released.

I considered the combination of input filtering and the architecture of the application. Both were pretty solid. I felt that the security posture was good enough with respect to addressing the minimum baseline. Now this task was completed. It was time to take a look at all security issues that have been created by the security squad in JIRA.

It quickly became clear that most of the tickets in JIRA were authorization issues. The same issues kept popping up all over the place. On top of that, we realized that the existing security tools were unable to identify these issues. Not even through customization. It simply wasn't possible.

"So how do we automate that? How do we turn these security issues into failing tests that pass once the fix is implemented?"

Testing was a serious activity in this project. We are talking about close to 100% code coverage across all microservices, end-to-end integration testing and a serious performance test regimen.

The only thing that was lacking was API-level integration tests for each endpoint in a microservice. Attempts were made by tech leads to achieve successful integration testing, but to no avail. There was nothing for us to add the security test cases to. The existing end-to-end integration testing was triggered from the frontend. It also went straight to the GraphQL server, which was not what we wanted the tests to run against. GraphQL is certainly great, but current security tools are not able to deal with it.

"Alright, so here we are now. How do we solve this?", I asked my team. "Why don't we create security unit tests?", someone replied. "Well, let's see. We have a dozen microservices and there are more coming all the time. Additionally, the functionality, especially on that level, is still changing quite often. I'm afraid we won't even be able to get proper security unit test coverage going. It's likely going to be a pain to maintain that and will cause more problems than it would solve. Just imagine the security unit tests start breaking builds left, right, and center."

Besides the obvious issues with that approach, we also wouldn't be able to run tests across the entire application. Also, we were limited to each individual microservice. We couldn't run tests for rate limiting, which was enforced on the API gateway. We couldn't ensure that the HTTP security headers stay in place. We couldn't even test the SSL configuration, nor leverage other dynamic security tools in that scenario. It was too narrow and brittle for what we wanted to achieve. And let's not even talk about maintaining that test suite.

We needed a solution that allowed us to run security integration tests from a central repository. This solution should also be able to run tests on different security aspects of the application. Including SSL configuration, API-endpoint security, and rate limiting. Additionally, I wanted to find a way to calculate security test coverage. I wanted to ensure that we had at least one security test for each endpoint, and could quickly get a list of new endpoints that were not covered yet.

As Lao Tzu famously said: "A journey of a thousand miles begins with a single step". I felt like our journey had only just begun.

Luckily, I remembered an open source security tool that I played around with a while ago. It is called BDD-Security and is a smart combination of Selenium, OWASP ZAP, and Cucumber. It leverages behavior-driven-development patterns and neatly ties it together into a security integration testing tool.

It supports many common security testing scenarios out of the box, including authentication and authorization based issues. After some minor customization, at least. It also wraps SSLyze, which allowed us to codify SSL related security expectations. BDD-Security seemed like it could be the perfect fit.

Diving right in, I started writing test cases for the first microservice. Yes, you thought right, the one from my first day.

The first challenge I encountered was due to the lack of documentation for covering web-services with BDD-Security. Much of the sample functionality relied on traditional web applications. After hacking away for a few hours I got most of the functionality working. At this stage, I had covered all endpoints with security tests and could run the test suite from the command line. It still had open issues, those were failing as expected. "That is pretty neat!", I mumbled to myself, "One down, 14 more to go".

Before I started with the other microservices I wanted to make my efforts visible to the team and embed it into the pipeline. Oh boy, if only I would have known the depth of the rabbit hole that I was getting myself into. This is a story for another time though. In short, it took more time than I want to admit of messing around with containerizing it and trying to make it work in the pipeline. A test-branch of a pipeline, that is. It wasn't quite ready for prime time. And to be frank, it wasn't very fast compared to the other test suites that were running in there.

Things didn't get any easier from there. Fast forward a couple of weeks and I'm still fighting to get all microservices covered. I've committed hundreds of lines of code, duplication is everywhere and the number of WTFs per minute increased with every hour I spent. Looking at it now, I've created a monster. A monster that is written in accordance to the "One-Factor App" methodology, also known as spaghetti code.

"That's a total write off.", I said to myself after a long, sad sigh. What I had created with the best of intentions was useless. It was time to move on.

Even though I stopped working on it, I still couldn't stop thinking about the problem. Then one day, by complete accident, I chanced upon something magnificent. The team had grown beyond 120 people by then. One of the recent newcomers had used Postman to create the integration test suite for a microservice. And yes, you are right - it was for that first ever microservice - my old nemesis. I had heard of Postman before, in fact, I had used it a few times. Mainly out of curiosity though. What I didn't know is that Postman comes with an API that supports pre-requests scripts, and fully fletched test cases. That was great, but the real kicker was Newman, the command line companion of Postman. Newman allows you to run any Postman collection via the command line interface, which means it runs nicely in pipelines.

It checked all of the boxes, and best of all, it was already being used by a squad. "This is going to catch on for sure and I'll finally be able to salvage my previous learnings. It was good for something, after all, I knew it!". I was beaming.

I immediately returned to my laptop, put on my headphones, tuned in to Brain.fm, cracked my knuckles and got started. It was one of those rare moments where everything just magically comes together. I made more progress in hours than I did in days before. I didn't have to write much code either. Except for the pre-request scripts that populate the environment with access tokens as well as the test cases that verify the expected behavior. It was all in there. It was a thing of beauty. I showed the progress to my team and they loved it. The monster that I had created earlier, couldn't be maintained by anyone but myself. This new set of Postman collections, however, allowed us to finally collaborate. Within a short amount of time, we were covering a decent percentage of the microservices. This was actually going to work out.

Because everything was going well, and the team did a lot of the heavy lifting, I became ambitious.

I was wondering, how could we measure security integration testing coverage? "Wouldn't it be great to know that all of the existing endpoints have security test cases? Wouldn't it be even better to know if there are any new endpoints that haven't been tested yet? A million times yes!" We were storing the Postman collections, which are JSON files. We had access to the swagger specifications, which are JSON files also. My spaghetti code senses started tingling again.

I spent the rest of that Wednesday afternoon creating a test coverage script. The code wasn't pretty but I was still proud of my achievement. The script would download the swagger specifications for all microservices that are deployed in SIT and UAT. Directly from Openshift. Next, it parsed the Postman collection and created a list of all endpoints that were called and how many times they were called. Finally, the swagger specifications were parsed. This produced the complete list of which endpoints are currently deployed. Also, it tracked how many security test requests each of them had. All of these changes were committed to source control. This would allow us to show a time series of how the endpoints of the entire app have been changing as well as how our test cases were gradually catching up with the deployed endpoints. Magnificent.

The central Postman collection had over 400 requests. Running Newman for the entire thing still took less than 5 minutes. We demoed it to a few people on several occasions and received a lot of positive feedback.

Our security squad felt great, this was clearly providing a lot of value. We didn't want to stop here. The self-proclaimed holy grail of security integration testing was the quest for solving one-time passwords within the test suite. And we were going for it. A Nexmo account with a local number was purchased, a Lambda function deployed to AWS and the Postman pre-request script coded. Et voila, we had cracked even that nut.

So let's recap what we had created. We were able to see which endpoints exist in all environments, and which of them had security tests. We were able to test all functionality across all environments. We could have easily created a subset of security integration tests to work as smoke tests for production. We were able to run these tests in the pipelines. The full test suite took less than 5 minutes to run. We could attach failing tests in a JIRA ticket to allow developers to quickly test their fixes. We created a dockerized version of this central script. This script could be generically added into the pipeline for every microservice repository. It was smart enough to understand which repository it was part of and run the right subfolder of the Postman collection. Oh yeah, and it solved freaking OTPs for you. *Boom, drop the mic*

But hang on for a moment. This is supposed be a story of Epic Failure, isn't it? So far it sounds like I've been nailing it. What's going on here?!

Don't worry, here it comes.

The Outcome

We were a security squad within this large development team. However, the security integration testing efforts happened in isolation. We were the only ones keeping it up to date, maintaining the 100% security test coverage. While there was a lot of good feedback, the feedback was not from the "right" people. I was caught up creating the perfect solution to a problem that I thought was critical. I followed the "build it and they will come" mantra.

I didn't stop and think about the future of this project. Remember, the team grew from one squad to a team of over 120. Half of the time I didn't even know who was a contractor, a consultant or an actual in-house developer. The team was so large and some excellent external consultants took the lead. I was busy making sure that they liked it. I didn't even stop and wonder who was going to carry the torch after that initial project milestone was hit. Now, almost a year later, the application is still there, but most of the external people that helped shape it are long gone.

Sure enough, our magnificent creation and all its potential for goodness never saw the light of day. That's a fail, but not of epic proportions. The sad thing is what happened to our central security integration repository. The one that we maintained for months. When the team switched to a new source code management system, it wasn't even copied over.

Understanding and respecting culture is the key to success in DevSecOps. And culture equals people. DevSecOps provides a unique opportunity for security engineers to make a difference. However, there is no room for heroes in DevSecOps.

Finding the right team members and obtaining key stakeholder commitment to security from the onset is more important than everything else. Look for them from the very beginning, then the rest will happen. Just start slow and take them on the journey.

"People don't resist change. They resist being changed." - Peter Senge.

Lessons Learned

- Don't waste time on over-engineering a security solution. Treat it as small experiments that have to be validated.
- No matter how great you think your solution is, it has to be built for the right people.
- Spend time on identifying influential people in the development team that can become the security champions. Tech leads are often great security champions.

Thank You

First of all, thanks to Edwin for coming up with the idea of writing a book on DevSecOps to help others by sharing our experiences.

Second, Mark without your support, expertise, and cat herding skills, this wouldn't have happened. Thanks a lot for all the help and the great execution of the book idea. I'm also thankful for all the authors that came together to create this book. Together we can reach so many more.

Last, but most definitely not least, my thanks go out to all the proofreaders that helped improve my chapter. Especially, Stephen Dye, Vicki Gatewood, Stephen McGowan, Simon Gerber, and Veronica Cryan.

About Stefan Streichsbier

Stefan began his career in Security Assurance in 2003 and has since performed intrusive security testing across hundreds of corporate networks and business-critical applications. Afterward, Stefan has been focused on secure application development for web and mobile applications, using his skills as both a developer and security expert to champion Source Code Analysis and Secure Application Coding best practice.

Stefan is regularly conducting security workshops, security awareness trainings, and frequently speaks at public events and conferences. Stefan has been dedicated to enabling organizations to rapidly deliver applications without creating a security bottleneck through application security programs and DevSecOps implementations.

Recently, Stefan founded GuardRails. A security platform that orchestrates open source security tools, curates their output and makes actionable results available in pull requests.

Stefan is a co-founder of the local DevSecOps Singapore Meetup group that is enjoying an active and ever-growing community. Stefan is also one of the core organizers of DevOpsDays Singapore, DevOpsDays Jakarta, and DevSecCon Asia.

Contact information:
- Twitter: @s_streichsbier
- Linkedin: sstreichsbier

Chapter 8

Strategic Asymmetry – Leveling the Playing Field Between Defenders and Adversaries

by Chetan Conikee

Chapter 8
Strategic Asymmetry – Leveling the Playing Field Between Defenders and Adversaries

Introduction

Understanding asymmetric conflict is key to building a successful #DevSecOps program.

In the early 70s, George Foreman was *the* undisputed heavyweight champion. None of his opponents had lasted more than three rounds in the ring and he was the strongest, hardest hitting boxer of his generation. Muhammad Ali, though not as powerful as Foreman, had a slightly faster punch and was lighter on his feet. Ali did not stand a chance against Foreman in the World Heavyweight Championship fight of October 1974. The outcome of that now-famous "**rumble in the jungle**" was completely unexpected. The two fighters were equally motivated to win. Both had boasted of victory, and both had enormous egos. Yet in the end, a fight that should have been over in three rounds went eight, and Foreman's prodigious punches proved useless against Ali's **rope-a-dope** strategy.

Foreman, confident of victory, pounded him again and again, while Ali whispered taunts: "*George, you're not hittin',*" "*George, you disappoint me.*" Foreman lost his temper, and his punches became a furious blur. By the fifth round, Foreman was worn out. And in round eight, as stunned commentators looked on, Ali knocked Foreman to the canvas, and the fight was long over.

The calculated endure-and-wait strategy of Ali versus Foreman's impulsive, unrelenting attacks can draw many parallels with modern cybersecurity world. In your cybersecurity initiative, would you consider yourself Ali or, a Foreman in this **asymmetric conflict**?

The Engineer manager would say - *"Yes, I am **Ali** in this **rumble** as I conduct static code analysis at a regular cadence. I measure the defect density across my entire application fleet on per build basis and have a tight control of my vulnerability programs."*

The #DevOps manager would say - *"Yes, I am **Ali** in this **rumble** as I have deployed bulletproof access management policies and a runtime security program that sandboxes, firewalls, and segments my workloads against bad actors."*

The Threat analyst would say - *"Yes, I am **Ali** in this **rumble** as I conduct adversarial threat modeling using honeypots and honeynets. I am a step ahead of this game"*

The CISO would say - *"Yes, I am **Ali** in this **rumble** as I aggregate my incidents in a SIEM (Security Information and Event Management) and throw some fairy dust of ML/AI/Deep Learning to yield actionable results"*

This fight illustrates an important yet relatively unexplored feature of conflict: how a weak actor's strategy can make a strong actor's power irrelevant. Phillips, A (2012) described attacks of this nature to be undetectable, and once occurred, impossible to determine its origin. This poses a fundamental question to security professionals:

How do we combat a threat we cannot see coming?

Compartmentalized thinking, guided by our confidence, expertise, bias, and heuristics puts us square in **Foreman**'s camp in this rumble as much as we think we are **Ali**.

"First-principles" thinking is one of the best ways to reverse-engineer this complicated problem and unleash creative possibility. Sometimes called, "reasoning from first principles," the idea is to break down complicated problems into basic elements and then reassemble them from the ground up. It's one of the best ways to unlock creative potential and move from linear to non-linear results.

The Story

Hardware is increasingly being abstracted away in favor of a more easily configurable, on-demand compute and storage (e.g., software-defined data centers). It is exclusively dependent on software running it - which takes the center stage in our times. Computer systems of today are increasingly built on SaaS models, as it can be developed, tested, and deployed more rapidly than managing and re-deploying hardware. This puts a large obligation on the quality of software.

Software systems can be protected by a multitude of mechanisms. For example, software that interacts with networks can be protected from unauthorized or malicious usage by firewalls, intrusion detection systems, or access control systems. In addition, software can be hardened against possible attacks by inspecting it, either statically or dynamically, to find security flaws during its development or in it's deployed state.

The security of a system can be characterized using the intrinsic relationship between four terms: *vulnerabilities, attacks, defenses,* and *policies.*

This relationship can be expressed in a sentence like, "*Defenses of type D enforce policies of class P to protect from attacks of class A exploiting vulnerabilities of class V*".

Let us deconstruct each of these terms to understand their relationship

Vulnerabilities *are defects or weaknesses in system security procedures, design, implementation, or internal controls that can be exercised and result in a security breach or violation of security policy.* - Gary McGraw, Software Security

Policies are the guarantees that a system can still give despite attacks. Conversely, we may state that some attack may break one or more policies. A customer relationship management system, for instance, enforces the policy that customer data is only disclosed to authenticated and authorized users of the system. Thus, the policy of the system expresses a confidentiality property on the customer data. Policies can express properties in either dimension of confidentiality, integrity, and availability.

Attacks are directed against a system's interface. The goal of an attack is always the infringement of at least one policy of a system. For instance, the infamous Distributed-Denial-of-Service (DDoS) attack flooding internet servers mostly target the availability of a system. However, even though the loss of availability is the most significant effect of such an attack, it might also be intended to break authentication systems to gain access to data, and thus, breaking the confidentiality of the system.

Defenses protect policies on systems from attacks. For instance, a firewall can protect a system from an unauthorized external access. At the heart of all defenses, there are three basic mechanisms: isolation, monitoring, and obfuscation. Isolation describes the physical or logical separation of a (sub)system's interfaces from each other or the environment. For example, a virtual machine is somewhat isolated from other virtual machines running on the same physical machine. Monitoring describes the observation of a system's interfaces and state in order to enforce **policies** as soon as a possible **violation** is detected or predicted as a consequence of known/unknown **vulnerability**. A firewall, for instance, observes the incoming and outgoing traffic of a network and, depending on the traffic's contents and the firewall ruleset decides to allow or block network traffic. Obfuscation protects data by making it unreadable without the knowledge of a secret. For instance, an encryption algorithm protects data from being understood without the knowledge of the correct key.

In the design of defenses and policies for systems, the *Principle of Least Privilege* should be the guiding factor. When systems are provided with policies and the defensive mechanisms to enforce them that limit the privilege of the system to the necessary minimum, attacks are less likely to succeed or to be harmful. These defensive mechanisms have to be carefully considered to enforce the policies in an effective way and can use isolation, monitoring, or obfuscation techniques or a combination of either of them.

With compartmentalized thinking, we often assume that focusing and tuning on one such concern is sufficient to define the overall security posture. We might even take it further by deploying instruments to measure all of these concerns in their respective silos. These instruments generate alerts which are then aggregated using a SIEM.

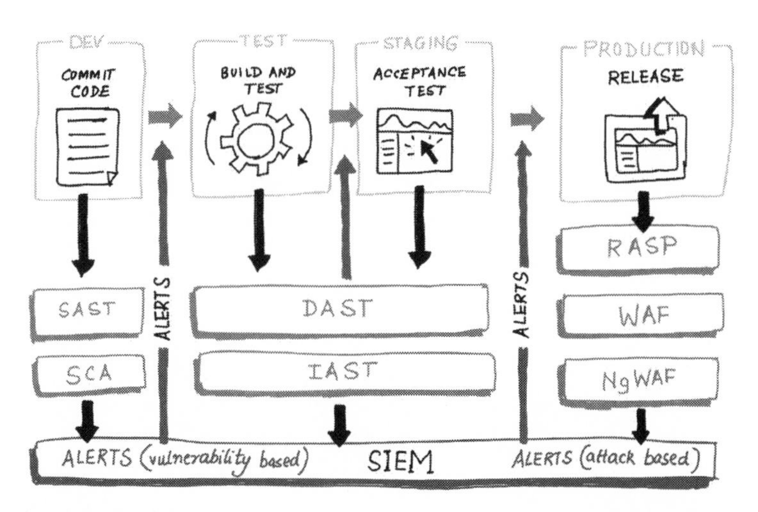

The promise of a SIEM is to bring context to your insight and make it actionable. You are now at the dispense of another instrument that can correlate data from these engines and create a narrative to act on.

The Outcome

Yet, the outcome has led to 159,000 cyber incidents, 7 billion records exposed, $5 billion financial impacts (**source**: One Trust Alliance report 2018). On a postmortem study, it was concluded that 93% of breaches could have been prevented.

Clearly, we've played directly to the asymmetry and advantage of an adversary.

Were these instruments ineffective?

Did we miss critical signals entangled in false positives?

Did our AI engine "CyberSec-HAL" fail us?

Lessons Learned

The Medieval castle approach inspired **defense in depth** thinking. Defense in Depth is the simple principle that while no security is perfect, the presence of many independent layers of defenses will increase the difficulty of an attacker to breach the walls and slow them down to the point where an attack isn't worth the expense it would take to initiate it. Defense in Depth places the core assets behind layers of varied and individually effective security, each of which has to be circumvented for an attack to be successful.

Applying defense in depth principles to our cloud instances leads to

- Conducting static (white-box) and dynamic (black-box) testing in earlier phases of life cycle to measure and control defect density.
- Hardening host operating system (disabling remote SSH to host or IP whitelisting access via jump host).
- Setting up a firewall and web application firewall. All incoming requests are assessed against a set of threat patterns to make "allow/deny" decision.

- Logging and Auditing across all layers of fabric to trace through adversarial behavior.
- Deploying honeypots and honeynets to model adversarial behavior.

Attacking applications deployed within your secure fabric is a classic case of **strategic asymmetry** as applications are portals to sensitive data and, unknowingly, to soft belly of the internal network.

Adversaries mainly use a broad "spray and pray" approach to opportunistically find targets, almost akin to a medieval catapult in front of a castle's drawbridge aiming at destroy siege towers and other siege engines of the attacking force.

Web applications and websites are the usual facades and front end of most businesses and organizations. In comparison to other hacking targets, they are easier to access and don't need any special connection or tools or state-sponsored resources, and when they're not intended to be used in an intranet, they can be accessed with any computer with an internet connection and a web browser. Exploiting common vulnerabilities like deserialization, command injection, XSS, and/or CSRF will be a trivial task and a walk in the park for script kiddies sitting in the comfort of their homes.

In many cases, once websites are breached, they serve as a beachhead for other major attacks and allow attackers to move laterally across the network with insider access, to escalate their privileges, and to eventually gain access to more critical resources such as databases, co-located applications, IAM keys, etc.

Does a disconnected/siloed defense in depth approach work?

Let us deconstruct instruments used in each silo and speak to its current state and what better can be done to created a connected fabric.

Silo # 1: Static and Dynamic testing to discover vulnerabilities in early phase of life cycle

Static analysis uses the foundation of syntax tree, control flow and dependence flow to generate alerts about potentially vulnerable conditions. However, the Intermediate Representation (IR) that is used to generate these alters are purged after a run cycle.

Discovering vulnerabilities can be likened to solving a puzzle. It comprises of identifying and thereafter enumerating all entry points i.e. a way a consumer would interact with the application programming interface (API). Each entry point can range from visibly modifiable parameters in the UI to interactions that are more obscure or transparent to the end-user. Each entry point when exercised would trigger a data flow comprising of a set of conditions required to serve a business need. Embedded within any of these flows are insecure states (vulnerabilities) that an adversary can manifest. Reachability defines how an adversary can use a modifiable parameter via an entry point to trigger an insecure state.

Connected rethink: It is thus imperative that an entry-exit point framework has a life beyond a single run cycle. This framework is akin to security-as-code as it is auto-generated by accessing software and defines the shape of an evolving application code in terms of its workflow and insecure states. With proper representation (Thrift, JSON, Protocol Buffers) it can be archived, version controlled and accessed to generate value elements. Using this framework one can evaluate it against a predefined policies to determine negative or positive drift.

Silo # 2: Protecting the application surface using a runtime agent or web application firewall

Web Application Firewalls (WAFs) are designed to inspect incoming traffic and use a set of signatures to infer intent. Depending on depth of configuration, it can be at worse overly permissive or at best overly protective. WAFs are primarily based on signatures (baselined from threat landscape) and are not tuned to adapt to application evolution. If not sustained, their efficacy will decay exponentially over time.

Typically, Runtime Application Security Protection (RASP) is instrumented with an application. When the application bootstraps itself in production, the RASP technique uses dynamic binary instrumentation or Byte-Code instrumentation (BCI) to add new security sensors and analysis capability to the entire application's surface. This process is very similar to how NewRelic or AppDynamics work to instrument an application for performance.

Upon instrumenting the entire surface, the agent can impose an inherent burden upon an application, further impacting both latency and throughput. These security sensors are tripped on every request in order to evaluate request metadata and other contextual information. If it looks like an attack, the request is tracked through the application. If the attack is causing the application to enter an inadmissible state (inferred from threat landscape or an adaptive learning system), it gets reported as a probe and the attack is blocked.

Connected rethink: Using this entry-exit framework proposed in **silo #1,** a baseline observation/protection policy can be defined to bootstrap

- Targeted red team attack strategy
- Auto-wired ruleset for a Web Application Firewall (WAF).
- Baseline policy for instrumented Runtime Protected Agent (RASP) to protect against known and unknown threats. The runtime agent monitors for imminent attack vectors leading to fundamental changes in observed behavior which in turn feeds back to augment policies.

Silo #3: Adversarial modeling using honeypots and honeynets

Honeypot Systems are decoy servers or systems set up to gather information regarding an attacker or intruder into your system. Honeypots can be set up inside, outside or in the DMZ of a firewall design or even in all of the locations although they are most often deployed inside of a firewall for control purposes. The deployment and usage of these tools are influenced by a number of technical and legal issues, which need to be carefully considered.

A honeynet consists of two or more honeypots on the same network. Honeypots may be deployed in combination like this to monitor larger or more diverse corporate networks and as part of a larger deception detection effort.

The Bitter Harvest paper presents a generic technique to systematically fingerprint low and medium at internet scale. Attackers have a strong motivation to detect honeypots at an early stage as they do not want to disclose their methods, exploits, and tools.

Connected rethink: There is still no such thing as an impenetrable system. Once attackers successfully breach a system, there is little to prevent them from doing arbitrary harm – but we can reduce the available **time** for the intruder to do this. But there can be the foundation for an "**immune system**" inspired approach to tackle zero-day and known exploits. Biological systems accept that defensive "walls" can be breached at several layers and therefore make use of an active and adaptive defense system to attack potential intruders - an **immune** system.

Modern, cloud-native platforms using distributed and elastic run-time environments, are provisioned using cloud formation templates. Software-defined provisioning enables rolling upgrades to entire fabric without downtimes.

An attacker passes through different stages to complete a cyber attack mission. It starts with initial **reconnaissance** and compromising of access means. The goal is to escalate privileges to get access to the target system by establishing a foothold near the system of interest. Attackers thereafter make use of counter-forensic measures to hide their presence and impair investigations. The next step is to conduct **lateral movement** to the target system. This is a complex and lengthy process and may even take weeks. The final step is **privilege escalation** leading to **exfiltration** or **collateral damage**.

Whenever an application is repositioned or redeployed (similar to rolling upgrades) all of its virtual machines are purged and regenerated. And this would effectively eliminate undetected hi-jacked machines. This makes it much harder for intruders to maintain a presence on victim systems which undergoes a purge process at random predefined intervals. The biological analogy of this strategy is called "**cell-regeneration**" and the attack on ill cells is coordinated by an immune system. This is an effective countermeasure – because the intruder immediately loses any hijacked machine albeit at which stage he might be in a cyber attack life cycle.

Such a biology-inspired immune system solution is charming but may also involve downsides. To regenerate too many nodes at the same time would let the system run "hot". This regeneration should be informed of fundamental changes in the observed behavior of application state.

Bringing it all together - One <u>Policy</u> to rule them all

No matter what the approach is, let us attempt to bring forth the spirit of DevOps to create an interconnected fabric. Let us collectively expand our thinking; guided by observation and systematic learning.

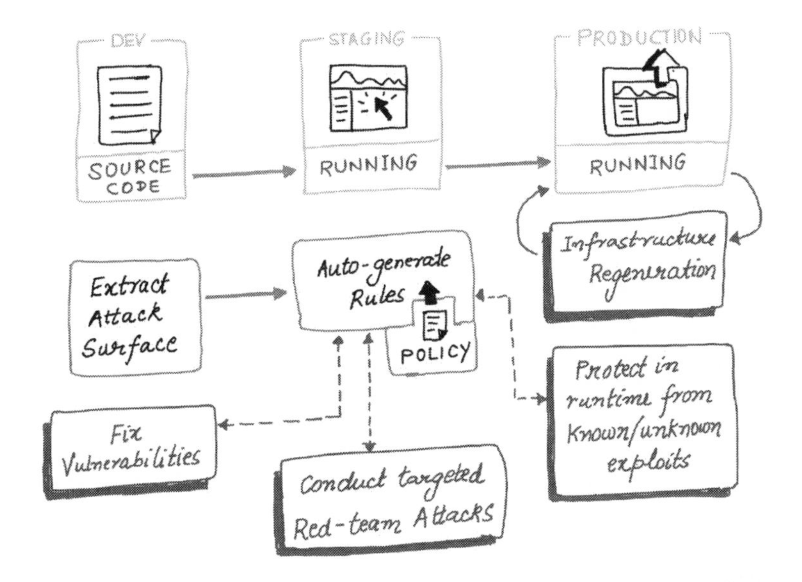

About Chetan Conikee

Chetan Conikee is a serial entrepreneur with over 20+ years of experience in authoring and architecting and securing mission-critical software. His expertise includes building web-scale distributed infrastructure, cybersecurity, personalization algorithms, complex event processing, fraud detection and prevention in investment/retail banking domains. He currently serves as CTO/Founder at ShiftLeft, and most recently Chief Data Officer and GM Operations at CloudPhysics.

Prior to CloudPhysics, Chetan was part of early founding teams at CashEdge (acquired FiServ), Business Signatures (acquired Entrust) and EndForce (acquired Sophos).

Conclusion

Conclusion

Malcolm Gladwell in his book "Outliers" talks about what it takes, the time and focus it takes, to internalize an idea to make it your own. Each of these authors has spent countless hours working and honing their craft. Much of that time was spent fixing failures. Our intent by presenting those failures in story form is to help you create your own narrative, not by copying our mistakes but by learning from them, and then creating your own failure tales.

Is there a shortcut to learning at this level? If we are to believe the research, no, not even those with innate talent can master their field without putting in the time and effort necessary to internalize the failures. According to the same research, the magic number to reach that level of mastery is 10,000 hours. We must suffer through years of frustration in order to grow and form our own, individual processes of learning.

Let's agree it's not necessary for us all to make the same mistakes, however. We each have to put in our time, but we don't have to all be breaking the same rock. Learn from others, share what you've learned and then, if all goes well, you'll have your own failures to brag about.

We encourage you to contribute to the DevSecOps Community with your own "Epic Failure". If we've done this properly, you understand the value of practitioner contributions and you'll be anxious to get started. We look forward to seeing your first story as part of the growing community at DevSecOpsDays.com.

Mark Miller
Founder and Editor in Chief, DevSecOpsDays.com
Co-Founder, All Day DevOps
Senior Storyteller, Sonatype

Acknowledgements

Acknowledgements

About DevSecOps Days Press

This is the first in a series of "Epic Failures" from DevSecOps Days Press. We'll continue to provide stories from people and teams throughout the community who want to contribute their story. Join us at DevSecOpsDays.com to find DevSecOps events in your area and to keep informed on upcoming projects, such as the DevSecOps Maturity Model, and the next book in this series, "Epic Failures in DevSecOps: Volume 2".

You're welcome to reach out to the authors for further discussions. They are all available on LinkedIn and are active community members in various forums.

Support for the Community

We couldn't have done this without the support of DevSecOpsDays.com and Sonatype. They have provided funding, resources and moral support, making it possible to create a community environment that will continue to grow as the community matures.

We invite you to join our community as a practitioner and as a contributor.

Thank You

This book is the work of eight author, but a lot went on behind the scenes to make all the pieces fit. We had over 150 people volunteer to proofread. As you can see at the conclusion of the chapters, the authors found the suggestions and comments from these volunteers invaluable to the refinement of their chapter.

The artwork for the project's book cover was provided by the design team at DevSecOpsDays.com. We have setup a community site that supports various aspects of the DevSecOps Community through articles, resources, podcasts and forums. Please join us as we continue to grow and act as a global hub for all things DevSecOps.

The "Epic Failures" series is a publication of DevSecOps Days Press with generous support from Sonatype.

Made in the USA
Columbia, SC
05 September 2019